APPROXIMATE
LINEAR ALGEBRAIC EQUATIONS

THE NEW UNIVERSITY MATHEMATICS SERIES

Editors: Professor E. T. Davies, *Department of Mathematics, University of Waterloo*, and Professor J. V. Armitage, *Shell Centre for Mathematical Education, University of Nottingham*

This series is intended for readers whose main interest is in mathematics, or who need the methods of mathematics in the study of science and technology. Some of the books will provide a sound treatment of topics essential in any mathematical training, while other, more advanced, volumes will be suitable as preliminary reading for research in the field covered. New titles will be added from time to time.

BRICKELL and CLARK: *Differentiable Manifolds: An Introduction*
BROWN and PAGE: *Elements of Functional Analysis*
BURGESS: *Analytical Topology*
COOPER: *Functions of a Real Variable*
CURLE and DAVIES: *Modern Fluid Dynamics* (Vols 1 and 2)
EASTHAM: *Theory of Ordinary Differential Equations*
KUPERMAN: *Approximate Linear Algebraic Equations*
MAUNDER: *Algebraic Topology*
PORTEOUS: *Topological Geometry*
ROACH: *Green's Functions: Introductory Theory with Applications*
RUND: *The Hamilton–Jacobi Theory in the Calculus of Variations*
SMITH: *Introduction to the Theory of Partial Differential Equations*
SMITH: *Laplace Transform Theory*
SPAIN: *Ordinary Differential Equations*
SPAIN: *Vector Analysis*
SPAIN and SMITH: *Functions of Mathematical Physics*
ZAMANSKY: *Linear Algebra and Analysis*

APPROXIMATE LINEAR ALGEBRAIC EQUATIONS

I. B. KUPERMAN

University of the Witwatersrand
Johannesburg

VAN NOSTRAND REINHOLD COMPANY

LONDON

NEW YORK   CINCINNATI   TORONTO   MELBOURNE

VAN NOSTRAND REINHOLD COMPANY LTD
Windsor House, 46 Victoria Street, London, S.W.1

INTERNATIONAL OFFICES
New York   Cincinnati   Toronto   Melbourne

Library of Congress Catalog Card No. 70–160200

ISBN 0 442 04546 8

First Published 1971

Paper supplied by
P. F. Bingham Limited, Croydon

Made and printed in Great Britain by
Butler and Tanner Ltd, Frome and London

Preface

Whenever the coefficients or right-hand constants of a system of linear algebraic equations are not known exactly, we have a system of *approximate linear algebraic equations.* This is the case, for example, when the coefficients and right-hand constants are obtained from measurement.

A number of methods are given in this book for obtaining the uncertainties in the unknowns due to the uncertainties in the coefficients and constants.

There is, however, no simple answer to the question: Which is the best, most practical, or recommended method? This depends on the magnitude of the given uncertainties, on the given coefficient matrix, on how accurately the uncertainties in the unknowns are required, and on the order of the given system of equations. For in certain cases, the volume of computation required to obtain the true intervals of uncertainty (Method IX) can become very large and indeed prohibitive. Under these circumstances, one must be satisfied with the results of methods leading to intervals containing the true intervals of uncertainty. Of these, Method VI or VIII leads to the best results, i.e., gives intervals of smallest width containing the true intervals of uncertainty.

In particular, it will be seen that the methods that give the best estimates of the true intervals of uncertainty require the most computation.

There is thus no simple answer as to which is the best method. But in many cases the statistical approach to approximate linear algebraic equations is the most appropriate (Chapter 12).

In order to make the book self-contained, certain mathematical topics with which all readers may not be familiar are dealt with briefly, namely, vector and matrix norms and the convergence of matrix series. Also, brief but adequate introductions are given to interval arithmetic, linear programming, and statistics, so as to make the important topic of approximate linear algebraic equations more easily accessible to a large readership.

Finally, let us say that the style was chosen so as to be best suited for the average reader on his first reading, bearing in mind that the book contains much source material.

Department of Applied Mathematics and Computer Science,
University of the Witwatersrand,
May, 1970   Israel B. Kuperman

Contents

To

RONA, AVRA and CHERYL

CHAPTER 1

Introduction

1.1 Introduction

We shall say that a system of linear algebraic equations is approximate if any or all of the coefficients and right-hand constants are not known exactly. If the values of the coefficients and right-hand constants depend upon physical measurements then clearly the values are not known exactly. If the measured values are expressed in decimal notation then it may safely be assumed that there is an uncertainty of at least one half-unit in the least significant position given, and usually by convention the uncertainty is one unit in the least significant position given. The uncertainties in the measured values may of course be considerably larger, but then the values of the uncertainties should be clearly stated. We call the study of the effects of the uncertainties in the coefficients and constants on the solution *approximate equation analysis.*

During the process of solution, the approximate equations are assumed to be exact, a solution having to be found which satisfies the accuracy requirements set, i.e., correct to the number of significant figures required. Solving a system of equations by two different methods or by using two different precisions, i.e., wordlengths, and comparing two such solutions may enable one to determine empirically the number of correct significant figures. Alternatively, interval arithmetic or theoretical error bounds may be used to determine the noise introduced by rounding. Thus, in solving the approximate system of equations, it is assumed that a sufficiently long wordlength is used, so that allowing for the effects of rounding noise leaves us with a solution correct at least to the required number of significant figures. Then, deciding to what accuracy the solution can be meaningfully used in view of the approximate nature of the equations is approximate equation analysis.

The uncertainties in the values of the unknowns are estimated from the uncertainties in the coefficients and constants. This facet of error analysis should form an integral part of a program for solving approximate linear algebraic equations and this is particularly important when small changes in the coefficients cause relatively large changes in the unknowns, i.e., in systems of equations which tend to become *ill-conditioned*, a term which we define later.

Now, it is well known that if we have a singular coefficient matrix, i.e., one whose determinant is zero, then the existence and nature of the solutions depend on whether the system is consistent or not. If the equations are inconsistent then there is no solution. And if the equations are consistent then there is an infinity of solutions, it being possible to choose the value of at least one of the unknowns arbitrarily. Thus, in a consistent system of equations with a singular coefficient matrix, at least one unknown can be chosen as large as we please in magnitude. We therefore assume the coefficient matrix of the approximate system of equations to be nonsingular, for otherwise the problem of finding a solution with finite uncertainties in the unknowns fails at the very beginning.

But suppose it is possible to find a singular coefficient matrix within the limits of the uncertainties in the coefficients. Then we say that such a system of equations is *critically ill-conditioned*; and in this case the true coefficient matrix may be singular within the limits of our knowledge.

Therefore, we ignore any solution of a critically ill-conditioned system of equations if a solution is sought with finite uncertainties and we say that for the given uncertainties in the coefficients no worthwhile solution can be found. We should then make quite sure that the physical situation giving rise to the equations can be expected to give n linearly independent equations in n unknowns. If this is the case, then a usable solution can only be obtained if the coefficients can be found more accurately, i.e., with smaller uncertainties.

1.2   A Critically Ill-conditioned System of Equations

As a numerical example of a system of equations

$$\mathbf{Ax} = \mathbf{c}, \quad \mathbf{A} = (a_{ij}), \quad \mathbf{x} = (x_j), \quad \mathbf{c} = (c_i),$$

in which there is an uncertainty of one unit in the least significant position given, consider the system of equations

$$\begin{aligned}
0{\cdot}974x_1 + 0{\cdot}790x_2 + 0{\cdot}311x_3 &= 2{\cdot}075 \\
-0{\cdot}631x_1 + 0{\cdot}470x_2 + 0{\cdot}251x_3 &= 0{\cdot}090 \\
0{\cdot}455x_1 + 0{\cdot}975x_2 + 0{\cdot}425x_3 &= 1{\cdot}855,
\end{aligned} \tag{1.1}$$

the uncertainty in each coefficient and right-hand constant being $0{\cdot}001$. The solution by Gaussian elimination (pivotal condensation) is given in Table 1.1 and is self-explanatory.

TABLE 1.1   *Solution of (1.1) by Gaussian Elimination†*

x_1	x_2	x_3	c	Row and explanation
0·974	0·790	0·311	2·075	R_1
−0·631	0·470	0·251	0·090	R_2
0·455	0·975	0·425	1·855	R_3
1	0·81109	0·31930	2·13039	$R_4 = R_1 \div 0·974$
	0·98180	0·45248	1·43428	$R_5 = R_2 + 0·631R_4$
	0·60595	0·27972	0·88567	$R_6 = R_3 - 0·455R_4$
	1	0·46087	1·46087	$R_7 = R_5 \div 0·98180$
		0·00045	0·00045	$R_8 = R_6 - 0·60595R_7$
		1	1·00000	$R_9 = R_8 \div 0·00045$
	1		1·00000	$R_{10} = R_7 - 0·46087R_9$
1			1·00000	$R_{11} = R_4 - 0·31930R_9 - 0·81109R_{10}$

† The solution is rounded to 5 decimal places, the solution having been obtained using a wordlength of 10 significant figures.

Referring to the table, we see that division of rows occurred on three occasions (rows R_4, R_7, and R_9). The divisors are called the *pivots* of the Gaussian elimination process, the three pivots in our case being 0·974, 0·98180, and 0·00045. Our procedure is to choose the coefficient of largest magnitude in the first column as the first pivot (underlined in row R_1) and to eliminate the unknown x_1 from the other equations, thereby obtaining the *reduced system of equations* in rows R_5 and R_6. Then we choose the coefficient of largest magnitude in the first column of the reduced system of equations (underlined in row R_5) and eliminate x_2 from the other equation in the reduced system of equations.

The third pivot is now 0·00045 in row R_8, and dividing by this pivot we obtain the value of x_3 in row R_9. This is the end of the so-called *forward procedure* of the Gaussian elimination process. And we may mention that the above method of choosing the pivot at each stage as the coefficient of largest magnitude in the first available column is called *partial pivoting*.

In rows R_{10} and R_{11} we in effect substitute the values of the unknowns obtained at each stage into previous equations until all the unknowns are found, the solution in our case being $x_1 = x_2 = x_3 = 1·00000$. This part of the solution is known as the *back-substitution procedure* of the Gaussian elimination process.

Now, it is instructive in our example to examine the effect of the uncertainty in a_{33} on the pivots. A little consideration of Table 1.1

shows that a change δa_{33} in the coefficient $a_{33} = 0.425$ changes the value of the last pivot 0.00045 in row R_8 by δa_{33} and leaves the other pivots unchanged.

Denoting the three pivots in rows R_1, R_5, and R_8 by p_1, p_2, and p_3, respectively, let us then consider the effect of a change δa_{33} given by

$$\delta a_{33} \in [-0.001, 0.001],$$

i.e., let us consider the effect of an uncertainty of 0.001 in a_{33}. Because a change δa_{33} in a_{33} produces a change δa_{33} in p_3 and leaves p_1 and p_2 unchanged, it follows that for

$$a_{33} \in [0.425 - 0.001, 0.425 + 0.001]$$

and all the other coefficients exact we have

$$p_3 \in [0.00045 - 0.001, 0.00045 + 0.001],$$

i.e.,

$$p_3 \in [-0.00055, 0.00145], \tag{1.2}$$

while $p_1 = 0.974$ and $p_2 = 0.98180$.

Thus, it is possible for the last pivot to be zero because the interval in (1.2) includes zero.

But in any Gaussian elimination process, the product of the pivots is, apart from sign, equal to the determinant of the coefficient matrix.

For one of the methods of evaluating a determinant is to reduce it by elementary row operations to the unit matrix, the determinant of the unit matrix being 1. And it may be recalled that in this procedure of evaluating determinants:

1. a P (permutation) elementary operation involves interchanging rows and changes the sign of the determinant,

2. an M (multiplication) elementary operation involves multiplying a row by a scalar and this multiplies the value of the determinant by the scalar, and

3. an A (addition) elementary operation involves adding to any one row multiples of any of the other rows; this does not change the value of the determinant.

But these elementary operations are involved in the Gaussian elimination process. The P elementary operation occurs if the coefficient chosen as pivot is not the first one in its column, and this then changes the sign of the determinant. The M elementary operation occurs when a pivotal row (i.e., a row containing a pivot) is divided by the pivot, the value of the determinant being divided by the pivot. And the A elementary operations occur during the elimination of the unknowns. These do not change the value of the determinant. Comparing Tables 1.1 and 1.2 may clarify the above. In Table 1.2, the whole system of equations is rewritten after each elementary operation, the original coefficient matrix being finally reduced by elementary operations to the unit matrix.

TABLE 1.2   *The Gaussian Elimination Process in Full for (1.1)†*

Coefficients			Constants	Corresponding operation in Table 1.1
0·974	0·790	0·311	2·075	R_1
−0·631	0·470	0·251	0·090	R_2
0·455	0·975	0·425	1·855	R_3
1	0·81109	0·31930	2·13039	$R_4 = R_1 \div 0·974$
−0·631	0·470	0·251	0·090	
0·455	0·975	0·425	1·855	
1	0·81109	0·31930	2·13039	
0	0·98180	0·45248	1·43428	$R_5 = R_2 + 0·631R_4$
0·455	0·975	0·425	1·855	
1	0·81109	0·31930	2·13039	
0	0·98180	0·45248	1·43428	
0	0·60595	0·27972	0·88567	$R_6 = R_3 − 0·455R_4$
1	0·81109	0·31930	2·13039	
0	1	0·46087	1·46087	$R_7 = R_5 \div 0·98180$
0	0·60595	0·27972	0·88567	
1	0·81109	0·31930	2·13039	
0	1	0·46087	1·46087	
0	0	0·00045	0·00045	$R_8 = R_6 − 0·60595R_7$
1	0·81109	0·31930	2·13039	
0	1	0·46087	1·46087	
0	0	1	1·00000	$R_9 = R_8 \div 0·00045$
1	0·81109	0·31930	2·13039	
0	1	0	1·00000	$R_{10} = R_7 − 0·46087R_9$
0	0	1	1·00000	
1	0	0	1·00000	$R_{11} = R_4 − 0·31930R_9 − 0·81109R_{10}$
0	1	0	1·00000	
0	0	1	1·00000	

† The system of equations is repeated after each elementary operation, showing clearly that, apart possibly from sign, the determinant of the coefficient matrix is equal to the product of the pivots. (In our example there is, however, no change of sign.)

Thus, in going from the given coefficient matrix to the unit matrix in Table 1.2, the value of the determinant is altered in magnitude whenever we divide by a pivot. Since the value of the determinant of the final unit matrix is unity, it follows that apart from sign (which depends on our choice of pivots) the value of the determinant of the coefficient matrix is equal to the product of the pivots.

It therefore follows that:

> If any of the pivots in a Gaussian elimination process can become zero within the limits of the uncertainties in the coefficients then the system of equations is critically ill-conditioned. (1.3)

In our example, the last pivot in Table 1.1 can certainly become zero for changes in the coefficients within the limits of their uncertainties; in fact, the last pivot can become zero within the limits of the uncertainty in a_{33} alone. Hence, the given system of equations in (1.1) is critically ill-conditioned for an uncertainty of $0 \cdot 001$ in each coefficient.

But it is not always possible to test for critical ill-conditioning by changing only one coefficient; simultaneous changes may have to be introduced in all the coefficients. We investigate this problem in Section 1.4, while we now introduce notation and state our problem more precisely in the next section.

1.3 Statement of Problem

Suppose that we are given a system of n linear algebraic equations in n unknowns

$$\mathbf{Ax} = \mathbf{c}, \quad \mathbf{A} = (a_{ij}), \quad \mathbf{x} = (x_i), \quad \mathbf{c} = (c_i) \qquad (1.4)$$

in which the coefficients a_{ij} and the right-hand constants c_i are the approximate values, the true values not being known exactly. Then, restricting ourselves to the case where $\mathbf{A}$ is nonsingular, the solution of (1.4) is

$$\mathbf{x} = \mathbf{Bc} \qquad (1.5)$$

where

$$\mathbf{B} = \mathbf{A}^{-1}, \quad \mathbf{B} = (b_{ij}).$$

If, in fact, the true system of equations corresponding to (1.4) is

$$\mathbf{A^*x^*} = \mathbf{c^*} \qquad (1.6)$$

let us suppose that the coefficients and right-hand constants of the true system of equations are known no more precisely than that given by

$$a_{ij}^* \in [a_{ij} - \varepsilon_{ij}, a_{ij} + \varepsilon_{ij}], \quad c_i^* \in [c_i - \varepsilon_i, c_i + \varepsilon_i],$$
$$i, j = 1, 2, \ldots, n, \qquad (1.7)$$

where the ε_{ij} and the ε_i are clearly nonnegative quantities. We call the ε_{ij} the *uncertainties in the coefficients* and the ε_i the *uncertainties in the right-hand constants*. And we call the intervals in (1.7) the *intervals of uncertainty in the coefficients* and the *intervals of uncertainty in the right-hand constants*, respectively.

We have thus chosen the approximate system of equations in (1.4) to correspond to the midpoints of the intervals in (1.7).

It may be pointed out that had the intervals of uncertainty in the coefficients and constants been given in the form

$$a_{ij}^* \in [f_{ij}, g_{ij}], \qquad c_i^* \in [u_i, v_i], \qquad i, j = 1, 2, \ldots, n, \qquad (1.8)$$

then, to correspond to the form in (1.7), we must take

$$a_{ij} = \tfrac{1}{2}(f_{ij} + g_{ij}), \qquad \varepsilon_{ij} = \tfrac{1}{2}(g_{ij} - f_{ij}), \qquad i, j = 1, 2, \ldots, n, \qquad (1.9)$$

and

$$c_i = \tfrac{1}{2}(u_i + v_i), \qquad \varepsilon_i = \tfrac{1}{2}(v_i - u_i), \qquad i = 1, 2, \ldots, n. \qquad (1.10)$$

Now, our first task in approximate equation analysis is clearly to satisfy ourselves that the true coefficient matrix $\mathbf{A}^*$ cannot become singular within the limits of the uncertainties ε_{ij}.

If, in fact, the approximate system of equations is not critically ill-conditioned, suppose that the x_i^* are given by

$$x_i^* \in [x_i - e_i, x_i + d_i], \qquad i = 1, 2, \ldots, n. \qquad (1.11)$$

And let us note that the d_i and e_i in the intervals are clearly nonnegative because one possible set of values of the x_i^* is $x_i^* = x_i$ $(i = 1, 2, \ldots, n)$ (see (9.2)).

Then, we call the intervals in (1.11) the *intervals of uncertainty in the unknowns* and we denote them by U_i $(i = 1, 2, \ldots, n)$, i.e.,

$$U_i = [x_i - e_i, x_i + d_i], \qquad i = 1, 2, \ldots, n. \qquad (1.12)$$

Now, the width or length of an interval $[a, b]$ is $(b - a)$ (see (9.4)). Hence the *widths of the intervals of uncertainty* denoted by

$$w(U_i), \qquad i = 1, 2, \ldots, n$$

are given by

$$w(U_i) = d_i + e_i, \qquad i = 1, 2, \ldots, n. \qquad (1.13)$$

And for each unknown x_i we call the larger of e_i and d_i in (1.11) the uncertainty Δx_i in the unknown, i.e., the *uncertainties in the unknowns* are

$$\Delta x_i = \max(e_i, d_i), \qquad i = 1, 2, \ldots, n. \qquad (1.14)$$

Clearly, we have by (1.12) and (1.14) that

$$U_i \subseteq [x_i - \Delta x_i, x_i + \Delta x_i], \qquad i = 1, 2, \ldots, n \qquad (1.15)$$

(see (9.3)). And hence

$$w(U_i) \leq w[x_i - \Delta x_i, \; x_i + \Delta x_i] = 2\Delta x_i, \qquad i = 1, 2, \ldots, n. \quad (1.16)$$

Alternatively, it follows directly from (1.13) and (1.14) that

$$w(U_i) \leq 2\Delta x_i, \qquad i = 1, 2, \ldots, n. \quad (1.17)$$

Our problem is thus to determine the intervals of uncertainty in the unknowns or the uncertainties in the unknowns for noncritically ill-conditioned systems of equations. A particular case of this we may point out is the determination of the intervals of uncertainty or the uncertainties in the elements of the inverse of an approximate coefficient matrix. For the elements of the inverse correspond to the unknowns when the right-hand columns of constants are the unit vectors (see Table 1.3).

And it should be pointed out that the theory developed can also be used in design problems. The coefficients and constants may be known exactly but the possible effects of varying these may have to be investigated.

1.4   The Ill-conditioning Factor

We now determine an approximate condition on critical ill-conditioning, defining an *ill-conditioning factor* ϕ. It will be seen that the necessary and sufficient condition for no critical ill-conditioning is $\phi < 1$. But certain second-order quantities are ignored in deriving this condition so that it is only an approximate condition (although a good one at that).

Let us first write the true system of equations in (1.6) $\mathbf{A^* x^*} = \mathbf{c^*}$ as

$$(\mathbf{A} + \delta\mathbf{A})(\mathbf{x} + \delta\mathbf{x}) = \mathbf{c} + \delta\mathbf{c}, \quad (1.18)$$

where $\mathbf{A}$, $\mathbf{x}$, and $\mathbf{c}$ refer to the approximate system of equations in (1.4). Further, let us write

$$\Delta\mathbf{A} = (\varepsilon_{ij})$$

and

$$|\delta\mathbf{A}| = (|\delta a_{ij}|), \qquad \text{where} \quad \delta\mathbf{A} = (\delta a_{ij}),$$

the modulus signs when applied to a matrix indicating that we are taking the absolute values of the elements of the matrix. And let us call $\Delta\mathbf{A}$ the *coefficient uncertainty matrix*.

Then, for no critical ill-conditioning of the approximate system of equations it is necessary and sufficient that $(\mathbf{A} + \delta\mathbf{A})$ be nonsingular for changes in the coefficients within the limits of the uncertainties, i.e., we must have that the determinant

$$\det(\mathbf{A} + \delta\mathbf{A}) \neq 0$$

whenever

$$|\delta\mathbf{A}| \leq \Delta\mathbf{A},$$

i.e., whenever

$$|\delta a_{ij}| \leq \varepsilon_{ij}, \qquad i, j = 1, 2, \ldots, n.$$

But the determinant is a function of the coefficients. Hence, writing

$$a = \det(\mathbf{A}) \qquad \text{and} \qquad a + \delta a = \det(\mathbf{A} + \delta\mathbf{A}),$$

the change δa in the determinant $a = \det(\mathbf{A})$ is given to the first order of small quantities by

$$\delta a = \sum_{i=1}^{n} \sum_{j=1}^{n} \frac{\partial a}{\partial a_{ij}} \delta a_{ij}. \tag{1.19}$$

Then, taking moduli we have that

$$|\delta a| \leq \sum_{i=1}^{n} \sum_{j=1}^{n} \left| \frac{\partial a}{\partial a_{ij}} \right| |\delta a_{ij}|. \tag{1.20}$$

And the equality sign is reached in (1.20), i.e.,

$$|\delta a| = \sum_{i=1}^{n} \sum_{j=1}^{n} \left| \frac{\partial a}{\partial a_{ij}} \right| |\delta a_{ij}|, \tag{1.21}$$

when $\partial a/\partial a_{ij}$ and δa_{ij} are always of the same sign or always of opposite sign in (1.19). For then each term in the double sum in (1.19) will be of the same sign.

Thus, for (1.21) to hold when the changes δa_{ij} are equal in magnitude to the uncertainties, i.e.,

$$|\delta a_{ij}| = \varepsilon_{ij}, \qquad i, j = 1, 2, \ldots, n, \tag{1.22}$$

we must have that either

$$\delta a_{ij} = \varepsilon_{ij} \operatorname{sign}\left(\frac{\partial a}{\partial a_{ij}}\right), \qquad \text{for all } i, j = 1, 2, \ldots, n, \tag{1.23}$$

or

$$\delta a_{ij} = -\varepsilon_{ij} \operatorname{sign}\left(\frac{\partial a}{\partial a_{ij}}\right), \qquad \text{for all } i, j = 1, 2, \ldots, n. \tag{1.24}$$

(The *sign function* is defined by

$$\operatorname{sign}(a) = 1, -1, 0 \qquad \text{according as } a > 0, \, a < 0, \, a = 0,$$

respectively.)

Let us now denote by Δa the least upper bound of the absolute change in value of the determinant within the limits of the uncertainties in the coefficients. Thus, within the limits of the uncertainties in the coefficients a change in value of the determinant as large as Δa in absolute value is possible, but a larger change in absolute value is not possible. (At this stage the reader has been introduced to $\delta\mathbf{A}$, $\Delta\mathbf{A}$, δa, and Δa, the change $\delta\mathbf{A}$ in the coefficient matrix $\mathbf{A}$ and the change δa in the determinant a of $\mathbf{A}$ being restricted by

$$|\delta\mathbf{A}| \leq \Delta\mathbf{A} = (\varepsilon_{ij}) \quad \text{and} \quad |\delta a| \leq \Delta a.)$$

Then, with the approximation leading to (1.19) we have that

$$\Delta a = \sum_{i=1}^{n} \sum_{j=1}^{n} \left|\frac{\partial a}{\partial a_{ij}}\right| \varepsilon_{ij}. \tag{1.25}$$

For, clearly, the expression on the right-hand side is an upper bound of $|\delta a|$ for

$$|\delta a_{ij}| \leq \varepsilon_{ij}, \quad i, j = 1, 2, \ldots, n. \tag{1.26}$$

And the expression on the right-hand side of (1.25) is in fact the least upper bound, because at least one set of changes δa_{ij} exists subject to (1.26) such that the maximum is reached, i.e.,

$$|\delta a| = \Delta a,$$

it being noted that in (1.19)

$$\delta a = \Delta a \text{ for the changes in (1.23)} \tag{1.27}$$

and

$$\delta a = -\Delta a \text{ for the changes in (1.24).} \tag{1.28}$$

Now, critical ill-conditioning implies that the determinant of the coefficient matrix can become zero within the limits of the uncertainties in the coefficients, and no critical ill-conditioning implies that the determinant cannot become zero within the limits of the uncertainties.

But by (1.27) and (1.28) the true value of the determinant is contained in

$$[a - \Delta a, a + \Delta a],$$

the necessary and sufficient condition for this interval not to contain zero being $\Delta a < |a|$, whatever the sign of a.

Hence, the necessary and sufficient condition for no critical ill-conditioning is

$$\Delta a < |a|, \quad \text{i.e.,} \quad \Delta a/|a| < 1,$$

i.e.,

$$\phi < 1 \tag{1.29}$$

where the *ill-conditioning factor* ϕ is defined by

$$\phi = \left(\sum_{i=1}^{n} \sum_{j=1}^{n} \left|\frac{\partial a}{\partial a_{ij}}\right| \varepsilon_{ij}\right) \Big/ |a| \tag{1.30}$$

(see (1.25)).

We have thus shown that the necessary and sufficient condition for no critical ill-conditioning is $\phi < 1$, i.e., we have shown that $\det(\mathbf{A} + \delta\mathbf{A}) \neq 0$ for $|\delta\mathbf{A}| \leq \Delta\mathbf{A}$ if and only if $\phi < 1$. But let us remember that in showing this we used an approximate relation in (1.19).

It may be noted that the ill-conditioning factor ϕ is clearly a non-negative quantity and that it is zero when the coefficients are exact, i.e., when all the $\varepsilon_{ij} = 0$.

1.5   The Evaluation of the Ill-conditioning Factor

We now consider the evaluation of the partial derivatives in the ill-conditioning factor in (1.30).

The expansion of a determinant in terms of the elements of its ith row is given by

$$a \equiv \det(\mathbf{A}) = \sum_{j=1}^{n} (-1)^{i+j} a_{ij} M_{ij} \qquad (1.31)$$

where M_{ij}, the i, jth minor of $\mathbf{A}$, is the determinant of the submatrix obtained by deleting the ith row and the jth column of the matrix $\mathbf{A}$.

(For example, suppose we evaluate the determinant of

$$\mathbf{A} = \begin{bmatrix} 0\text{·}53 & 0\text{·}86 & 0\text{·}48 \\ 0\text{·}94 & -0\text{·}47 & 0\text{·}85 \\ 0\text{·}87 & 0\text{·}55 & 0\text{·}26 \end{bmatrix}$$

in terms of the elements of the second row. Then,

$$a = \det(\mathbf{A}) = (-1)^{2+1} a_{21} M_{21} + (-1)^{2+2} a_{22} M_{22} + (-1)^{2+3} a_{23} M_{23}$$

$$= -0\text{·}94 \times \begin{vmatrix} 0\text{·}86 & 0\text{·}48 \\ 0\text{·}55 & 0\text{·}26 \end{vmatrix} - 0\text{·}47 \times \begin{vmatrix} 0\text{·}53 & 0\text{·}48 \\ 0\text{·}87 & 0\text{·}26 \end{vmatrix}$$

$$- 0\text{·}85 \times \begin{vmatrix} 0\text{·}53 & 0\text{·}86 \\ 0\text{·}87 & 0\text{·}55 \end{vmatrix}$$

$$= -0\text{·}94 \times 0\text{·}0404 + 0\text{·}47 \times 0\text{·}2798 - 0\text{·}85 \times 0\text{·}4567$$
$$= 0\text{·}5577,$$

i.e.,

$$a = 0\text{·}5577. \qquad (1.32)$$

And this result, of course, agrees with the value of the determinant obtained by any other method, as, for example, by finding the product of the three pivots $0\text{·}94$, $1\text{·}1250$, and $-0\text{·}5274$ in Table 1.3 and then changing the sign of this product because the first pivot is not in the first row.)

Now, on differentiating the n terms in (1.31) partially with respect to a_{ik} we obtain

$$\frac{\partial a}{\partial a_{ik}} = (-1)^{i+k} M_{ik}, \qquad i, k = 1, 2, \ldots, n$$

since the minors M_{ij} in (1.31) ($j = 1, 2, \ldots, n$ and i fixed) contain no element of the ith row of $\mathbf{A}$ so that in particular they do not contain a_{ik}. In fact, only the term $(-1)^{i+k} a_{ik} M_{ik}$ in the sum in (1.31) contributes to the result. Thus, with a change of indices we have

$$\frac{\partial a}{\partial a_{ij}} = (-1)^{i+j} M_{ij}, \qquad i, j = 1, 2, \ldots, n. \tag{1.33}$$

But by a well-known result in determinant theory

$$(-1)^{i+j} M_{ij}/a$$

are the elements of the transpose of the inverse of $\mathbf{A}$, i.e.,

$$b_{ji} = (-1)^{i+j} M_{ij}/a \qquad i, j = 1, 2, \ldots, n$$

where $\mathbf{B} = (b_{ij}) = \mathbf{A}^{-1}$. Hence, by (1.33),

$$\frac{\partial a}{\partial a_{ij}} \Big/ a = b_{ji}, \qquad i, j = 1, 2, \ldots, n. \tag{1.34}$$

Therefore,

$$\left| \frac{\partial a}{\partial a_{ij}} \right| \Big/ |a| = |b_{ji}|, \qquad i, j = 1, 2, \ldots, n,$$

so that (1.30) becomes

$$\phi = \sum_{i=1}^{n} \sum_{j=1}^{n} |b_{ji}| \varepsilon_{ij}. \tag{1.35}$$

Thus, the evaluation of ϕ requires the inverse of the coefficient matrix. In the case where all the uncertainties ε_{ij} are equal to ε, the ill-conditioning factor takes the form

$$\phi = \left(\sum_{i=1}^{n} \sum_{j=1}^{n} |b_{ij}| \right) \varepsilon, \tag{1.36}$$

i.e., the ill-conditioning factor ϕ is then equal to the sum of the absolute values of the elements of the inverse multiplied by the uncertainty ε.

1.6   The Ill-conditioning Matrix C

We find it convenient to define an *ill-conditioning matrix* $\mathbf{C}$ by

$$\mathbf{C} = |\mathbf{B}| \Delta \mathbf{A} \qquad \text{where} \quad |\mathbf{B}| = (|b_{ij}|), \tag{1.37}$$

i.e., $\mathbf{C} = (c_{ij})$ where

$$c_{ij} = \sum_{k=1}^{n} |b_{ik}| \varepsilon_{kj}, \qquad i, j = 1, 2, \ldots, n. \tag{1.38}$$

We now first express the ill-conditioning factor ϕ in terms of the elements of the ill-conditioning matrix $\mathbf{C}$ and then derive some properties of $\mathbf{C}$ when all the $\varepsilon_{ij} = \varepsilon$.

Now, by (1.38) we have

$$\sum_{i=1}^{n} c_{ii} = \sum_{i=1}^{n} \sum_{k=1}^{n} |b_{ik}| \varepsilon_{ki} = \sum_{j=1}^{n} \sum_{i=1}^{n} |b_{ji}| \varepsilon_{ij}, \quad \text{on changing dummy indices,}$$

$$= \sum_{i=1}^{n} \sum_{j=1}^{n} |b_{ji}| \varepsilon_{ij}, \quad \text{on changing the order of}$$

summation. Hence,

$$\phi = \sum_{i=1}^{n} c_{ii} \tag{1.39}$$

by (1.35). Thus, the ill-conditioning factor ϕ is equal to the sum of the diagonal elements of the ill-conditioning matrix $\mathbf{C}$.

And when all the $\varepsilon_{ij} = \varepsilon$, we have by (1.38) that

$$c_{kj} = \sum_{i=1}^{n} |b_{ki}| \varepsilon_{ij} = \varepsilon \sum_{i=1}^{n} |b_{ki}| = b_k \varepsilon \tag{1.40}$$

where

$$b_k = \sum_{i=1}^{n} |b_{ki}| = \text{sum of absolute values of elements in } k\text{th}$$
$$\text{row of inverse.} \tag{1.41}$$

Thus:

> When all the $\varepsilon_{ij} = \varepsilon$ the columns of $\mathbf{C}$ are all equal, the elements in the ith row of $\mathbf{C}$ being by (1.40) and (1.41) equal to ε multiplied by the sum of the absolute values of the elements in the ith row of the inverse of $\mathbf{A}$. (1.42)

1.7 Examples

Example 1

We now determine the ill-conditioning factor ϕ for the system of equations

TABLE 1.3   Solution of (1.43) and Inverse of Coefficient Matrix by Gaussian Elimination†

x_1	x_2	x_3	c	e_1	e_2	e_3	Row	Explanation
0·53	0·86	0·48	0·64	1	0	0	R_1	
0·94	−0·47	0·85	3·12	0	1	0	R_2	
0·87	0·55	0·26	0·83	0	0	1	R_3	
1	−0·5000	0·9043	3·3191	0	1·0638	0	R_4	$R_2 \div 0·94$
	1·1250	0·0007	−1·1191	1	−0·5638	0	R_5	$R_1 − 0·53R_4$
	0·9850	−0·5267	−2·0577	0	−0·9255	1	R_6	$R_3 − 0·87R_4$
	1	0·0007	−0·9948	0·8889	−0·5012	0	R_7	$R_5 \div 1·1250$
		−0·5274	−1·0778	−0·8756	−0·4319	1	R_8	$R_6 − 0·9850R_7$
		1	2·0438	1·6603	0·8189	−1·8963	R_9	$R_8 \div (−0·5274)$
			−0·9962	0·8878	−0·5017	0·0013	R_{10}	$R_7 − 0·0007R_9$
1			0·9730	−1·0574	0·0724	1·7153	R_{11}	$R_4 + 0·5000R_{10} − 0·9043R_9$
1			0·9730	−1·0574	0·0724	1·7153	R_{12}	$R_{12} = R_{11}$
	1		−0·9962	0·8878	−0·5017	0·0013	R_{13}	$R_{13} = R_{10}$
		1	2·0438	1·6603	0·8189	−1·8963	R_{14}	$R_{14} = R_9$

† The true values have been rounded to 4D (4 decimal places). The coefficient matrix in rows R_1, R_2, and R_3 has been transformed by elementary row operations to the unit matrix in rows R_{12}, R_{13}, and R_{14}, the operations in rows R_{12} to R_{14} being equivalent to reversing the order of rows R_9 to R_{11}.

$$0.53x_1 + 0.86x_2 + 0.48x_3 = 0.64$$
$$0.94x_1 - 0.47x_2 + 0.85x_3 = 3.12 \tag{1.43}$$
$$0.87x_1 + 0.55x_2 + 0.26x_3 = 0.83$$

for

$$\varepsilon_{ij} = 0.01 = \varepsilon_i, \qquad i, j = 1, 2, 3. \tag{1.44}$$

The solution of the system of equations in (1.43) and the inverse of the coefficient matrix

$$\mathbf{A} = \begin{bmatrix} 0.53 & 0.86 & 0.48 \\ 0.94 & -0.47 & 0.85 \\ 0.87 & 0.55 & 0.26 \end{bmatrix} \tag{1.45}$$

are found in Table 1.3.

The inverse is obtained in Table 1.3 by the technique of including columns of constants corresponding to the unit vectors and constituting the unit matrix, this technique being justified in the next paragraph.

It is well known that the effect of carrying out a series of elementary row operations on a given matrix is equivalent to premultiplying it by a suitable matrix. In Table 1.3 the coefficient matrix $\mathbf{A}$, say, in the first three rows and the first three columns is transformed to the unit matrix $\mathbf{I}$ in the last three rows (and first three columns). This must be equivalent to premultiplying $\mathbf{A}$ by its inverse $\mathbf{B}$ because the inverse of $\mathbf{A}$ is uniquely given by $\mathbf{BA} = \mathbf{I}$. But the unit matrix in the first three rows in the columns headed e_1, e_2, and e_3 undergoes the same elementary operations as $\mathbf{A}$ so that the matrix in the last three rows in the columns headed e_1, e_2, and e_3 is $\mathbf{BI}$. Therefore, since $\mathbf{BI} = \mathbf{B}$ it follows that the matrix in the last three rows in Table 1.3 in the columns headed e_1, e_2, and e_3 is the inverse of the coefficient matrix $\mathbf{A}$.

Thus, from Table 1.3 the inverse of the coefficient matrix is

$$\mathbf{A}^{-1} \equiv \mathbf{B} = \begin{bmatrix} -1.0574 & 0.0724 & 1.7153 \\ 0.8878 & -0.5017 & 0.0013 \\ 1.6603 & 0.8189 & -1.8963 \end{bmatrix} \tag{1.46}$$

and the solution of the system of equations to 4D is

$$x_1 = 0.9730, \qquad x_2 = -0.9962, \qquad x_3 = 2.0438. \tag{1.47}$$

Hence, for the uncertainties in (1.44) we have by (1.36) that

$$\phi = 0.01 \times (1.0574 + 0.0724 + 1.7153 + 0.8878 + 0.5017$$
$$+ 0.0013 + 1.6603 + 0.8189 + 1.8963),$$

i.e.,

$$\phi = 0.08611. \tag{1.48}$$

Thus, the system of equations in (1.43) is not critically ill-conditioned for the uncertainties in (1.44), it being noted that the values of the ε_{ij} in (1.44) and not those of the ε_i are involved in determining ϕ.

At this stage, it may be instructive to determine the ill-conditioning matrix $\mathbf{C}$ in (1.37) for our example and to determine ϕ by (1.39).

By (1.40) and (1.46) we have

$$c_{11} = c_{12} = c_{13} = b_1\varepsilon = (|-1{\cdot}0574| + |0{\cdot}0724| + |1{\cdot}7153|) \times 0{\cdot}01$$
$$= 0{\cdot}028451.$$

Similarly,

$$c_{2j} = 0{\cdot}013908, \qquad c_{3j} = 0{\cdot}043755, \qquad j = 1, 2, 3.$$

Thus,

$$\mathbf{C} = \begin{bmatrix} 0{\cdot}028451 & 0{\cdot}028451 & 0{\cdot}028451 \\ 0{\cdot}013908 & 0{\cdot}013908 & 0{\cdot}013908 \\ 0{\cdot}043755 & 0{\cdot}043755 & 0{\cdot}043755 \end{bmatrix}. \tag{1.49}$$

We note that the columns of $\mathbf{C}$ are identical because we have an example here where all the $\varepsilon_{ij} = \varepsilon$ (see (1.42)).

And by (1.49) and (1.39) we have

$$\phi = 0{\cdot}028451 + 0{\cdot}013908 + 0{\cdot}043755 = 0{\cdot}08611 \tag{1.50}$$

which is, of course, the same result as in (1.48).

Example 2

By way of a further example, let us determine the value of ε for the system of equations in (1.1) to become critically ill-conditioned.

The inverse of the coefficient matrix can be shown to be

$$\mathbf{B} = \begin{bmatrix} 43{\cdot}03 & 31{\cdot}81 & 50{\cdot}20 \\ 373{\cdot}55 & 266{\cdot}47 & 430{\cdot}10 \\ 810{\cdot}89 & 577{\cdot}26 & 935{\cdot}30 \end{bmatrix}. \tag{1.51}$$

Hence, by (1.36) and (1.29) the value of ε for no critical ill-conditioning is given by

$$(43{\cdot}03 + 31{\cdot}81 + 50{\cdot}20 + 373{\cdot}55 + 266{\cdot}47 + 430{\cdot}10$$
$$+ 810{\cdot}89 + 577{\cdot}26 + 935{\cdot}30)\varepsilon < 1,$$

i.e.,

$$\varepsilon < 0{\cdot}000284. \tag{1.52}$$

Now, it is clear by (1.36) and (1.51) that the small value of ε for the system of equations to become critically ill-conditioned is due to the large values of the elements of the inverse of the coefficient matrix.

Suppose, now, that by saying that a system of equations has a *tendency to ill-conditioning* we shall imply that the system of equations becomes critically ill-conditioned for small values of the uncertainties in the coefficients. Then, clearly, a system of equations has a tendency to ill-conditioning if the elements of the inverse are large. But care must be taken in using the term, as will be seen in the next section.

1.8 Effects of Scaling

We now deal briefly with the effects of scaling a given system of equations on $\mathbf{C}$, ϕ, and on the use of the term *tendency to ill-conditioning*. And we also consider the effects of changing the order of the equations in the system.

By scaling, we mean that the various equations in the system are divided by arbitrary nonzero scalars so as to produce a system of equations more convenient for some purpose. (This may be required, for example, if the Jacobi or Gauss–Seidel iterative processes are used to solve the system of equations.)

It should be clear in the first place that neither scaling nor changing the order of the equations affects the solution.

Now, suppose the kth equation is divided by a constant s. This will cause each of the uncertainties in this equation to be divided by s. Thus, the uncertainties in the coefficients in the kth row of the coefficient matrix are divided by s.

And it is easy to see that the effect of this scaling is for the elements in the kth column of the inverse to be multiplied by s. For, to find the inverse of the scaled coefficient matrix, suppose we use the technique of introducing a unit matrix, as in Table 1.3. Then, let the first operation be to multiply the kth row of the scaled coefficient matrix and of the unit matrix by s. We then obtain our original coefficient matrix, but the element in the kth row in the column headed e_k is now s instead of unity. Hence, the inverse of the scaled coefficient matrix will be the same as for the original coefficient matrix, except that the elements in the kth column are all multiplied by s.

It follows that the effect of dividing the kth equation by s is to leave the elements of $\mathbf{C}$ unaltered. For, in evaluating

$$c_{ij} = \sum_{h=1}^{n} |b_{ih}| \varepsilon_{hj}, \qquad i, j = 1, 2, \ldots, n,$$

each term is unaltered, for with $h = k$ the uncertainty ε_{kj} is divided by s while the other factor $|b_{ik}|$ is multiplied by s.

And it is easy to see that simultaneously scaling all the equations in a given system of equations will leave the elements of $\mathbf{C}$ unaltered.

Now, since the ill-conditioning factor ϕ can be expressed in terms of the elements of $\mathbf{C}$ (see (1.39)), it also follows that scaling does not alter ϕ.

Next, let us consider the effect of changing the order of two equations, say the pth and qth equations. This will have the effect of interchanging the ε_{pj} and ε_{qj}, $j = 1, 2, \ldots, n$ (and also, of course, of interchanging the uncertainties ε_p and ε_q in the right-hand constants). And it is well known that the effect of interchanging the pth and qth rows of the co-efficient matrix is to interchange the pth and qth columns of the inverse, i.e., to interchange b_{ip} and b_{iq}, $i = 1, 2, \ldots, n$.

The permutation will, however, not alter the elements of $\mathbf{C}$:

$$c_{ij} = \sum_{k=1}^{n} |b_{ik}| \varepsilon_{kj}, \qquad i, j = 1, 2, \ldots, n,$$

the pth and qth terms in the sums being merely interchanged.

And it is easy to see that the effect of any permutation of the equations will leave the elements of $\mathbf{C}$ unaltered, which is the same as saying that the matrix $\mathbf{C}$ is left unaltered. This result must also hold for the ill-conditioning factor ϕ.

Combining our results we then have:

> The ill-conditioning matrix $\mathbf{C}$ and the ill-conditioning factor ϕ are not changed by any scaling of the system of equations or by any change of order of the equations in the system. $\hspace{1em}$ (1.53)

Further, we have seen that the effect on the inverse of the coefficient matrix is as follows:

> Scaling a system of equations such that the ith equation, $i = 1, 2, \ldots, n$, is divided by a scalar k_i will cause the corresponding column of the inverse to be multiplied by k_i. And any permutation of the equations will produce the corresponding permutation on the columns of the inverse. $\hspace{1em}$ (1.54)

Since in applying the term *tendency to ill-conditioning* we are interested in the magnitude of the elements of the inverse, we have only to allow for the effects of scaling. Let us then define the following standard form for a system of equations:

> We say that a system of equations is scaled in *standard form* if the coefficient of largest magnitude in each equation is unity. $\hspace{1em}$ (1.55)

Thus, when applying the qualitative term *tendency to ill-conditioning* let us suppose that the system of equations is first scaled in standard form. Then, if the elements of the inverse are large so that the system will become critically ill-conditioned for small values of the uncertainties we shall say that the system has a tendency to ill-conditioning.

We begin dealing with the determination of the uncertainties in the unknowns in Chapter 3. But first, in Chapter 2, after dealing with some mathematical preliminaries, we derive certain strict sufficient conditions for no critical ill-conditioning, it being recalled that the condition $\phi < 1$ in (1.29) involves the approximation made in obtaining (1.19). (It will be seen, however, that $\phi < 1$ is strictly a sufficient condition for no critical ill-conditioning when all the $\varepsilon_{ij} = \varepsilon$.)

CHAPTER 2

Sufficient Conditions for No Critical Ill-conditioning

The reader familiar with the mathematical preliminaries in Sections 2.1 to 2.3 should go directly to Sections 2.4, where we derive sufficient conditions for no critical ill-conditioning.

2.1 Vector and Matrix Norms

Vector and matrix norms are measures associated with vectors and matrices subject to certain conditions. But before stating the conditions which a measure must satisfy to be classed a norm, let us give an example of a vector norm and of a matrix norm. In considering the conditions the reader may then check whether our examples indeed satisfy these.

Given a vector $\mathbf{x} = (x_i)$ the absolute value of the element of maximum magnitude is one vector norm, i.e.,

$$\max_i(|x_i|) \tag{2.1}$$

is a vector norm. And given a square matrix $\mathbf{A} = (a_{ij})$ of order n one matrix norm is

$$\max_i\left(\sum_{j=1}^{n}|a_{ij}|\right), \tag{2.2}$$

i.e., the maximum row sum of the absolute values of the elements is a matrix norm.

In general, the *norm of a vector* $\mathbf{x}$ is a nonnegative number denoted by $\|\mathbf{x}\|$ corresponding to $\mathbf{x}$, which satisfies the following conditions:

(a) $\|\mathbf{x}\| > 0$ if $\mathbf{x} \neq \mathbf{0}$, and $\|\mathbf{0}\| = 0$;

(b) $\|c\mathbf{x}\| = |c|\,\|\mathbf{x}\|$ for any scalar c;   (2.3)

(c) $\|\mathbf{x} + \mathbf{y}\| \leq \|\mathbf{x}\| + \|\mathbf{y}\|$, the 'triangle inequality'.

Clearly, the measure in (2.1) is a vector norm because conditions (a) to (c) above are satisfied.

For example, suppose $\mathbf{x} = (1, -3, 4)$, $\mathbf{y} = (5, 2, 4)$, so that $\mathbf{x} + \mathbf{y} = (6, -1, 8)$. Then, for the vector norm defined in (2.1) condition (c) in (2.3) is clearly satisfied:

$$\|\mathbf{x} + \mathbf{y}\| = 8, \qquad \|\mathbf{x}\| = 4, \qquad \|\mathbf{y}\| = 5.$$

Subscripts are frequently used to distinguish between different vector norms, the so-called first and second vector norms of a vector $\mathbf{x}$ of dimension n being assigned as follows:

$$\|\mathbf{x}\|_{\mathrm{I}} = \max_i(|x_i|) \tag{2.4}$$

$$\|\mathbf{x}\|_{\mathrm{II}} = |x_1| + |x_2| + \ldots + |x_n| = \sum_{i=1}^{n} |x_i|. \tag{2.5}$$

For the norms defined above, condition (c) in (2.3) is clearly satisfied, because for the components of any two vectors $\mathbf{x} = (x_i)$ and $\mathbf{y} = (y_i)$ of dimension n we have $|x_i + y_i| \leq |x_i| + |y_i|$ $(i = 1, 2, \ldots, n)$.

Next, we consider matrix norms.

In general, the *norm of a square matrix* $\mathbf{A}$ is a nonnegative number denoted by $\|\mathbf{A}\|$ corresponding to $\mathbf{A}$ which satisfies the following conditions:

(a) $\|\mathbf{A}\| > 0$ if $\mathbf{A} \neq \mathbf{0}$, and $\|\mathbf{0}\| = 0$;

(b) $\|c\mathbf{A}\| = |c| \, \|\mathbf{A}\|$ for any scalar c;   (2.6)

(c) $\|\mathbf{A} + \mathbf{B}\| \leq \|\mathbf{A}\| + \|\mathbf{B}\|$;

(d) $\|\mathbf{A}\mathbf{B}\| \leq \|\mathbf{A}\| \, \|\mathbf{B}\|$.

It is easy to check that conditions (a) to (c) are satisfied by the measure in (2.2) and condition (d) is in fact also satisfied; that the measure in (2.2) is a matrix norm will become clear later, after dealing with the so-called compatibility condition.

If

$$\|\mathbf{A}\mathbf{x}\| \leq \|\mathbf{A}\| \, \|\mathbf{x}\| \tag{2.7}$$

for certain procedures of assigning matrix and vector norms for any matrix $\mathbf{A}$ and any vector $\mathbf{x}$ (for which the vector $\mathbf{A}\mathbf{x}$ exists), then the matrix norm in question is said to be *compatible* with the vector norm being used. And such a matrix norm (i.e., one that is compatible with some vector norm) is sometimes called a *natural* matrix norm.

Now, given a particular procedure of assigning vector norms, suppose we determine the maximum of the vector norms $\|\mathbf{A}\mathbf{x}\|$ for a given matrix $\mathbf{A}$ on the supposition that $\mathbf{x}$ runs over the set of all vectors for which $\|\mathbf{x}\| = 1$, i.e., suppose we determine

$$\max_{\|x\|=1}(\|\mathbf{Ax}\|).$$

And suppose we associate this measure with $\mathbf{A}$. Then it can be shown that this measure satisfies conditions (a) to (d) in (2.6) for matrix norms and also that the compatibility relation (2.7) is satisfied (see reference 1, pp. 56–59).

Thus:

$$\|\mathbf{A}\| = \max_{\|x\|=1}(\|\mathbf{Ax}\|) \text{ is a matrix norm satisfying the}$$
$$\text{compatibility relation } \|\mathbf{Ax}\| \leq \|\mathbf{A}\|\,\|\mathbf{x}\|. \tag{2.8}$$

And a matrix norm constructed in the manner indicated in (2.8) is said to be *subordinate* to the vector norm in question.

Let us now construct the matrix norm subordinate to the first vector norm $\|\mathbf{x}\|_I$ in (2.4) and let us denote this norm by $\|\mathbf{A}\|_I$.

Then, by (2.8)

$$\|\mathbf{A}\|_I = \max_{\|x\|_I\,=1}(\|\mathbf{Ax}\|_I), \tag{2.9}$$

and it can be shown that

$$\|\mathbf{A}\|_I = \max_i\left(\sum_{j=1}^{n}|a_{ij}|\right) \tag{2.10}$$
$$= \text{maximum row sum of absolute values.}$$

Thus, the measure in (2.2) is a matrix norm, i.e., it satisfies conditions (a) to (d) in (2.6), and, further, it is compatible with the first vector norm, i.e.,

$$\|\mathbf{Ax}\|_I \leq \|\mathbf{A}\|_I\,\|\mathbf{x}\|_I.$$

And the above result of course holds for any vector $\mathbf{x}$ and any matrix $\mathbf{A}$ (see (2.7)).

For example, for the matrix

$$\mathbf{A} = \begin{bmatrix} 3 & 4 & 5 \\ 2 & -6 & 8 \\ 2 & 1 & 0 \end{bmatrix} \tag{2.11}$$

we have $\|\mathbf{A}\|_I = |2| + |-6| + |8| = 16,$

the choice of $\mathbf{x}$ to achieve the maximum in (2.9) in this case being the column vector $(1, -1, 1)'$ or $(-1, 1, -1)'$, the dash to indicate transpose.

Further, suppose we are given a vector $\mathbf{x} = (2, 3, -4)'$. Then for $\mathbf{A}$ in (2.11) it is easy to check that $\mathbf{Ax} = (-2, -46, 7)'$. Thus,

$$\|\mathbf{Ax}\|_I = 46, \qquad \|\mathbf{A}\|_I = 16, \qquad \|\mathbf{x}\|_I = 4, \tag{2.12}$$

so that the compatibility relation $\|\mathbf{Ax}\| \leq \|\mathbf{A}\|\,\|\mathbf{x}\|$ in (2.7) is satisfied for the example.

And if we are given that

$$\mathbf{A} = \begin{bmatrix} 3 & 4 & 5 \\ 2 & -6 & 8 \\ 2 & 1 & 0 \end{bmatrix}, \qquad \mathbf{B} = \begin{bmatrix} 1 & -2 & 0 \\ 0 & 1 & 0 \\ 3 & 0 & 1 \end{bmatrix},$$

so that

$$\mathbf{AB} = \begin{bmatrix} 18 & -2 & 5 \\ 26 & -10 & 8 \\ 2 & -3 & 0 \end{bmatrix} \tag{2.13}$$

it is easy to check that condition (d) in (2.6) is satisfied for the example. For, clearly,

$$\|\mathbf{AB}\|_{\mathrm{I}} = |26| + |-10| + |8| = 44,$$
$$\|\mathbf{A}\|_{\mathrm{I}} = |2| + |-6| + |8| = 16,$$
$$\|\mathbf{B}\|_{\mathrm{I}} = |3| + |0| + |1| = 4$$

so that $\|\mathbf{AB}\|_{\mathrm{I}} \leq \|\mathbf{A}\|_{\mathrm{I}}\,\|\mathbf{B}\|_{\mathrm{I}}$ for our example.

Now, let us construct the matrix norm compatible with the second vector norm $\|\mathbf{x}\|_{\mathrm{II}}$ in (2.5), and let us denote this matrix norm by $\|\mathbf{A}\|_{\mathrm{II}}$. Then, by (2.8),

$$\|\mathbf{A}\|_{\mathrm{II}} = \max_{\|x\|_{\mathrm{II}} = 1} (\|\mathbf{Ax}\|_{\mathrm{II}}) \tag{2.14}$$

and it can be shown that

$$\|\mathbf{A}\|_{\mathrm{II}} = \max_j \left(\sum_{i=1}^{n} |a_{ij}| \right) \tag{2.15}$$
$$= \text{maximum column sum of absolute values.}$$

For the matrix $\mathbf{A}$ in (2.11) we thus have

$$\|\mathbf{A}\|_{\mathrm{II}} = |5| + |8| + |0| = 13,$$

the choice of $\mathbf{x}$ to achieve the maximum in (2.14) in our example being the column vector $(0, 0, 1)'$ or $(0, 0, -1)'$.

And corresponding to (2.12) we have for the same $\mathbf{x}$, $\mathbf{A}$, and hence $\mathbf{Ax}$ (i.e., for $\mathbf{x} = (2, 3, -4)'$, $\mathbf{A}$ in (2.11), and $\mathbf{Ax} = (-2, -46, 7)'$) that

$$\|\mathbf{Ax}\|_{\mathrm{II}} = |-2| + |-46| + |7| = 55,$$
$$\|\mathbf{A}\|_{\mathrm{II}} = |5| + |8| + |0| = 13,$$
$$\|\mathbf{x}\|_{\mathrm{II}} = |2| + |3| + |-4| = 9.$$

Thus, the compatibility relation $\|\mathbf{Ax}\|_{\mathrm{II}} \leq \|\mathbf{A}\|_{\mathrm{II}}\,\|\mathbf{x}\|_{\mathrm{II}}$ is satisfied in this example.

Further, for the matrices $\mathbf{A}$ and $\mathbf{B}$ in (2.13) we have that

$$\|\mathbf{AB}\|_{\mathrm{II}} = |18| + |26| + |2| = 46,$$
$$\|\mathbf{A}\|_{\mathrm{II}} = |5| + |8| + |0| = 13,$$
$$\|\mathbf{B}\|_{\mathrm{II}} = |1| + |0| + |3| = 4$$

so that condition (d) in (2.6) is satisfied in the example, i.e.,

$$\|\mathbf{AB}\|_{II} \leq \|\mathbf{A}\|_{II} \|\mathbf{B}\|_{II}.$$

Summarizing, then, we have introduced in this section the vector norms $\|\mathbf{x}\|_I$ and $\|\mathbf{x}\|_{II}$ and the matrix norms $\|\mathbf{A}\|_I$ and $\|\mathbf{A}\|_{II}$. These satisfy the conditions in (2.3) and (2.6) and further the compatibility relation (2.7) holds for $\|\mathbf{A}\|_I$ and $\|\mathbf{x}\|_I$ and for $\|\mathbf{A}\|_{II}$ and $\|\mathbf{x}\|_{II}$, i.e.,

$$\|\mathbf{Ax}\|_I \leq \|\mathbf{A}\|_I \|\mathbf{x}\|_I$$

and

$$\|\mathbf{Ax}\|_{II} \leq \|\mathbf{A}\|_{II} \|\mathbf{x}\|_{II}. \tag{2.16}$$

2.2 Convergent Matrices

A square matrix $\mathbf{A}$ is said to be *convergent* if

$$\lim_{m \to \infty} \mathbf{A}^m = \mathbf{0}, \tag{2.17}$$

i.e., if the limit as m approaches infinity of each element of the mth power of $\mathbf{A}$ is zero.

Now, the necessary and sufficient condition for a matrix $\mathbf{A}$ to be convergent involves the eigenvalues of $\mathbf{A}$, so we briefly recall their definition and some results, defining the term *spectral radius* in (2.22).

The eigenvalues of a square matrix $\mathbf{A}$ are those values of λ for which nontrivial solutions of $\mathbf{x}$ exist in $\mathbf{Ax} = \lambda\mathbf{x}$, i.e., in

$$(\mathbf{A} - \lambda\mathbf{I})\mathbf{x} = \mathbf{0}, \tag{2.18}$$

a nontrivial solution being one other than $\mathbf{x} = \mathbf{0}$.

But it is well known that a homogeneous system of equations has nontrivial solutions only if the determinant of the coefficient matrix is zero. It follows by (2.18) that the eigenvalues satisfy the equation

$$\det(\mathbf{A} - \lambda\mathbf{I}) = 0, \tag{2.19}$$

i.e.,

$$\begin{vmatrix} a_{11} - \lambda & a_{12} & \ldots a_{1n} \\ a_{21} & a_{22} - \lambda & \ldots a_{2n} \\ \ldots & & \\ a_{n1} & a_{n2} & \ldots a_{nn} - \lambda \end{vmatrix} = 0. \tag{2.20}$$

This is a polynomial equation of degree n in λ, it has n roots, and these are the n eigenvalues of $\mathbf{A}$:

$$\lambda_1, \lambda_2, \ldots, \lambda_n. \tag{2.21}$$

In general, the eigenvalues are complex and may be repeated, the absolute value of the eigenvalue of maximum magnitude being called the *spectral radius* of the matrix $\mathbf{A}$ and being denoted by $\rho(\mathbf{A})$, i.e.,

$$\rho(\mathbf{A}) = \max_i(|\lambda_i|). \tag{2.22}$$

We now state without proof (see reference 1, p. 60) the necessary and sufficient condition for a square matrix $\mathbf{A}$ to be convergent:

> A square matrix $\mathbf{A}$ is convergent if and only if $\rho(\mathbf{A}) < 1$.   (2.23)

But a well-known result in the theory of norms (see reference 1 p. 61) is:

> The spectral radius can exceed no matrix norm, i.e., $\rho(\mathbf{A}) \leq \|\mathbf{A}\|$ whatever matrix norm is employed.   (2.24)

It follows that if $\|\mathbf{A}\| < 1$ then $\rho(\mathbf{A}) < 1$ so that by (2.23) we have the following useful sufficient condition for a matrix $\mathbf{A}$ to be convergent:

> A matrix $\mathbf{A}$ is convergent if any one of its norms is less than unity, i.e.,
> $$\mathbf{A}^m \rightarrow \mathbf{0} \quad \text{as} \quad m \rightarrow \infty \quad \text{if} \quad \|\mathbf{A}\| < 1 \qquad (2.25)$$
> for any one matrix norm.

2.3 Convergence of Matrix Series

We now consider the convergence of the matrix series

$$\mathbf{I} + \mathbf{A} + \mathbf{A}^2 + \ldots + \mathbf{A}^m + \ldots . \qquad (2.26)$$

It can be shown that:

> The series $\mathbf{I} + \mathbf{A} + \mathbf{A}^2 + \ldots + \mathbf{A}^m + \ldots$ converges if and only if $\mathbf{A}$ is convergent, the sum of the series being then equal to $(\mathbf{I} - \mathbf{A})^{-1}$.   (2.27)

In particular, we may note that:

> If the series $\mathbf{I} + \mathbf{A} + \mathbf{A}^2 + \ldots + \mathbf{A}^m \ldots$ converges its sum is equal to $(\mathbf{I} - \mathbf{A})^{-1}$.   (2.28)

And it follows by (2.27) and (2.23) that:

> The series $\mathbf{I} + \mathbf{A} + \mathbf{A}^2 + \ldots + \mathbf{A}^m + \ldots$ converges if and only if $\rho(\mathbf{A}) < 1$.   (2.29)

Then, by (2.24), we have the following useful sufficient condition for the convergence of the matrix series:

> The series $\mathbf{I} + \mathbf{A} + \mathbf{A}^2 + \ldots + \mathbf{A}^m + \ldots$ converges if $\|\mathbf{A}\| < 1$ for any one matrix norm.   (2.30)

We now consider the errors introduced when the series (2.26) converges if its summation is terminated at the term $\mathbf{A}^k$.

By (2.28), the error matrix due to the summation of the series (2.26) being terminated at $\mathbf{A}^k$ is clearly

$$(\mathbf{I} - \mathbf{A})^{-1} - (\mathbf{I} + \mathbf{A} + \ldots + \mathbf{A}^k) = \mathbf{A}^{k+1} + \mathbf{A}^{k+2} + \ldots.$$

Hence, taking norms, we have

$$\|(\mathbf{I} - \mathbf{A})^{-1} - (\mathbf{I} + \mathbf{A} + \mathbf{A}^2 + \ldots + \mathbf{A}^k)\| \leq \|\mathbf{A}^{k+1}\| + \|\mathbf{A}^{k+2}\| + \ldots$$
$$\leq \|\mathbf{A}\|^{k+1} + \|\mathbf{A}\|^{k+2} + \ldots$$
$$= \|\mathbf{A}\|^{k+1} (1 + \|\mathbf{A}\| + \ldots).$$

Hence,

$$\|(\mathbf{I} - \mathbf{A})^{-1} - (\mathbf{I} + \mathbf{A} + \mathbf{A}^2 + \ldots + \mathbf{A}^k)\| \leq \frac{\|\mathbf{A}\|^{k+1}}{1 - \|\mathbf{A}\|} \quad (2.31)$$

if $\qquad\qquad\qquad \|\mathbf{A}\| < 1,$

it being recalled that the sum of the geometric series $1 + r + r^2 + \ldots$ is $1/(1 - r)$ if $|r| < 1$.

This, then, gives a bound for the norm of the error matrix introduced by terminating the series in (2.26) at the term $\mathbf{A}^k$.

2.4 Sufficient Conditions for No Critical Ill-conditioning

It is easy to see that a sufficient condition for a matrix $(\mathbf{S} + \mathbf{I})$ to be nonsingular is $\rho(\mathbf{S}) < 1$, where the spectral radius $\rho(\mathbf{S})$ of $\mathbf{S}$ is by definition equal to the absolute value of the eigenvalue of maximum magnitude. For the eigenvalues of $\mathbf{S}$ are those values of λ for which $\det(\mathbf{S} - \lambda\mathbf{I}) = 0$, so that if $\max(|\lambda_i|) \equiv \rho(\mathbf{S}) < 1$ then $\det(\mathbf{S} + \mathbf{I}) \neq 0$. For, if $\det(\mathbf{S} + \mathbf{I}) = 0$, then $\lambda = -1$, so that $\rho(\mathbf{S}) = \max(|\lambda_i|) \not< 1$.

Thus

$$\det(\mathbf{S} + \mathbf{I}) \neq 0 \quad \text{if} \quad \rho(\mathbf{S}) < 1. \quad (2.32)$$

Now, writing

$$\mathbf{A} + \delta\mathbf{A} = \mathbf{A}(\mathbf{I} + \mathbf{B}\delta\mathbf{A})$$

and taking determinants, we have

$$\det(\mathbf{A} + \delta\mathbf{A}) = \det(\mathbf{A}) \det(\mathbf{I} + \mathbf{B}\delta\mathbf{A}),$$

it being recalled that the determinant of the product of two matrices is equal to the product of the determinants. Hence, $\det(\mathbf{A} + \delta\mathbf{A}) \neq 0$ if and only if $\det(\mathbf{I} + \mathbf{B}\delta\mathbf{A}) \neq 0$, because $\det(\mathbf{A}) \neq 0$ by assumption. It follows, by (2.32), that $\mathbf{A} + \delta\mathbf{A}$ is nonsingular if $\rho(\mathbf{B}\delta\mathbf{A}) < 1$, i.e.,

$$\det(\mathbf{A} + \delta\mathbf{A}) \neq 0 \quad \text{if} \quad \rho(\mathbf{B}\delta\mathbf{A}) < 1. \quad (2.33)$$

But a well-known result in matrix theory states that for any two square matrices $\mathbf{F} = (f_{ij})$ and $\mathbf{G} = (g_{ij})$ of order n

$$\rho(\mathbf{F}) \leq \rho(\mathbf{G}) \quad \text{if} \quad |f_{ij}| \leq g_{ij}, \quad i, j = 1, 2, \ldots, n \quad (2.34)$$

where the elements of $\mathbf{F}$ may be complex and where $\mathbf{G}$ is clearly a nonnegative matrix (see reference 2, p. 289). Hence,

$$\rho(\mathbf{B}\delta\mathbf{A}) \leq \rho(|\mathbf{B}|\Delta\mathbf{A}) = \rho(\mathbf{C}), \qquad \text{when} \quad |\delta\mathbf{A}| \leq \Delta\mathbf{A}, \qquad (2.35)$$

because then we clearly have that $|\mathbf{B}\delta\mathbf{A}| \leq |\mathbf{B}|\Delta\mathbf{A}$.

It follows from (2.35) that

$$\rho(\mathbf{B}\delta\mathbf{A}) < 1 \qquad \text{if} \quad \rho(\mathbf{C}) < 1 \qquad \text{for} \quad |\delta\mathbf{A}| \leq \Delta\mathbf{A}.$$

Hence, by (2.33),

$$\det(\mathbf{A} + \delta\mathbf{A}) \neq 0 \qquad \text{for} \quad |\delta\mathbf{A}| \leq \Delta\mathbf{A} \qquad \text{if } \rho(\mathbf{C}) < 1.$$

Thus $\mathbf{A} + \delta\mathbf{A}$ is nonsingular for $\delta\mathbf{A}$ within the limits of the uncertainties if $\rho(\mathbf{C}) < 1$.

But by (2.24) the spectral radius does not exceed any of the matrix norms, i.e., $\rho(\mathbf{C}) \leq \|\mathbf{C}\|$ whatever matrix norm is employed.

Thus:

A sufficient condition for no critical ill-conditioning is $\rho(\mathbf{C}) < 1$, and this is satisfied if $\|\mathbf{C}\| < 1$ for any one matrix norm.   (2.36)

In particular, choosing first and second matrix norms and writing

$$\phi_{\text{norm}} \equiv \|\mathbf{C}\|_{\mathrm{I}} \qquad (2.37)$$

we have by (2.36) that

$$\phi_{\text{norm}} = \|\mathbf{C}\|_{\mathrm{I}} < 1 \qquad (2.38)$$

or

$$\|\mathbf{C}\|_{\mathrm{II}} < 1 \qquad (2.39)$$

is a sufficient condition for no critical ill-conditioning.

While we may of course at all times use the stronger condition $\rho(\mathbf{C}) < 1$ in (2.36) in place of the conditions in (2.38) and (2.39), we should note that the first and second matrix norms are generally easier to compute than the spectral radius. Further, the weaker conditions in (2.38) and (2.39) may be expected to be sufficiently good in practice. We therefore pay due attention to them.

2.5 Special Case

We now consider the forms taken by the sufficient conditions (2.38), (2.39), and (2.36) for no critical ill-conditioning when all the $\varepsilon_{ij} = \varepsilon$.

In (2.38) we have by (1.40) that

$$\phi_{\text{norm}} = \|\mathbf{C}\|_{\mathrm{I}} = \max_k \left(\sum_{j=1}^{n} |c_{kj}| \right) = \max_k (b_k) n\varepsilon = \|\mathbf{B}\|_{\mathrm{I}} n\varepsilon,$$

i.e.,

$$\phi_{\text{norm}} = \|\mathbf{B}\|_{\mathrm{I}} n\varepsilon. \qquad (2.40)$$

Thus, by (2.38) and (2.40) a sufficient condition for no critical ill-conditioning when all the $\varepsilon_{ij} = \varepsilon$ is

$$\phi_{\text{norm}} = \|\mathbf{B}\|_1 n\varepsilon < 1. \tag{2.41}$$

And in (2.39) we have by (1.40) that

$$\|\mathbf{C}\|_{\text{II}} = \max_j \left(\sum_{k=1}^{n} |c_{kj}| \right) = \max_j \left(\varepsilon \sum_{k=}^{n} |b_k| \right),$$

i.e.,

$$\|\mathbf{C}\|_{\text{II}} = \left(\sum_{i=1}^{n} \sum_{j=1}^{n} |b_{ij}| \right) \varepsilon = \phi \tag{2.42}$$

(see (1.36)).

It thus follows by (2.39) and (2.42) that:

> When all the $\varepsilon_{ij} = \varepsilon$ then $\phi < 1$ is strictly a sufficient condition for no critical ill-conditioning, $\tag{2.43}$

the above derivation involving no approximations, unlike that for (1.29).

And now let us consider the sufficient condition $\rho(\mathbf{C}) < 1$ for no critical ill-conditioning for our special case when all the $\varepsilon_{ij} = \varepsilon$ (see (2.36)).

By definition in (1.41), the row sums b_k are nonnegative. And further, none of the b_k can be zero, for otherwise the elements in the corresponding row of **B** would all be zero so that **B** would be singular. Hence, for $\varepsilon \neq 0$ it is clear by (1.40) that **C** is a positive matrix (i.e., one with all its elements greater than zero). But it is well known that a positive matrix has a simple positive eigenvalue which is greater than the absolute values of all the other eigenvalues and which is not less than the smallest column sum nor greater than the largest column sum (see reference 3, p. 3). But by (1.42) all the column sums of **C** are equal when all the $\varepsilon_{ij} = \varepsilon$. Hence $\rho(\mathbf{C}) = \|\mathbf{C}\|_{\text{II}}$. And for the trivial case $\varepsilon = 0$ we have by (1.40) that $\mathbf{C} = \mathbf{0}$, so that we again have that $\rho(\mathbf{C}) = \|\mathbf{C}\|_{\text{II}}$, each side being equal to zero.

Thus:

$$\rho(\mathbf{C}) = \|\mathbf{C}\|_{\text{II}}, \qquad \text{when all } \varepsilon_{ij} = \varepsilon. \tag{2.44}$$

Hence, by (2.42),

$$\rho(\mathbf{C}) = \|\mathbf{C}\|_{\text{II}} = \phi, \qquad \text{when all } \varepsilon_{ij} = \varepsilon. \tag{2.45}$$

Hence, for our special case the sufficient condition $\rho(\mathbf{C}) < 1$ for no critical ill-conditioning in (2.36) becomes:

> When all the $\varepsilon_{ij} = \varepsilon$ then $\rho(\mathbf{C}) = \|\mathbf{C}\|_{\text{II}} = \phi < 1$ is a sufficient condition for no critical ill-conditioning. $\tag{2.46}$

We may now note that it follows from (2.40) and (2.42) that

$$\phi_{\text{norm}} = \|\mathbf{B}\|_1 n\varepsilon \geq \sum_{i=1}^{n} \sum_{j=1}^{n} |b_{ij}|\varepsilon = \phi.$$

For, clearly, $\|\mathbf{B}\|_1 n$, i.e., n times the maximum row sum, cannot be less than the n row sums, i.e., cannot be less than the sum of the absolute values of all the elements of $\mathbf{B}$. Thus,

$$\phi_{\text{norm}} \geq \phi, \qquad \text{when all } \varepsilon_{ij} = \varepsilon. \tag{2.47}$$

And this result could also have been derived from (2.45), because by (2.24) $\rho(\mathbf{C}) \leq \|\mathbf{C}\|$ whatever norm is employed so that

$$\rho(\mathbf{C}) \leq \|\mathbf{C}\|_1 = \phi_{\text{norm}}.$$

Clearly, then, (2.47) follows from (2.45). Thus, ϕ_{norm} can be regarded as a pessimistic estimate of ϕ for the case when all the $\varepsilon_{ij} = \varepsilon$.

For the numerical example in (1.43) and (1.44) we have by (2.38) and (1.49) that

$$\phi_{\text{norm}} \equiv \|\mathbf{C}\|_1 = |0\cdot043755| + |0\cdot043755| + |0\cdot043755| = 0\cdot1313, \tag{2.48}$$

while by (1.48) or (1.50) we have that $\phi = 0\cdot08611$. Thus these results illustrate (2.47).

And by (2.45) we have that

$$\rho(\mathbf{C}) = 0\cdot08611 \tag{2.49}$$

for the example.

Having derived the sufficient conditions (2.36), (2.38), and (2.39) for no critical ill-conditioning, we deal next with the problem of finding the uncertainties in the unknowns.

References

1. FADDEEVA, V. N. *Computational Methods of Linear Algebra*, Dover, New York (1959).
2. TAUSSKY, O. 'Some topics concerning bounds for eigenvalues of finite matrices', *Survey in Numerical Analysis* (Ed. John Todd), 279–97, McGraw-Hill, New York (1962).
3. BRAUER, A. 'On the characteristic roots of nonnegative matrices', *Recent Advances in Matrix Theory* (Ed. Hans Schneider), 3–38, University of Wisconsin Press, Madison and Milwaukee (1964).

CHAPTER 3

Method I: Upper Bounds of Uncertainties by Norm Analysis

3.1 Introduction

In this chapter we describe the simplest procedure for obtaining an upper bound for the uncertainties in the unknowns, the inverse of the coefficient matrix being required. This bound, which is in fact an upper bound of the largest uncertainty, is usually far larger than the largest uncertainty and no further information is available about the individual uncertainties. Nevertheless, the value indicated for the largest uncertainty may be small enough for our purpose, i.e., when the solution of the system of equations is required to a sufficiently limited accuracy. Then none of the other methods given in this book need be used.

But the condition of application of Method I is only a sufficient condition for no critical ill-conditioning. Hence, Method I may fail to give finite upper bounds for the uncertainties although there is no critical ill-conditioning, i.e., although the uncertainties in the unknowns are finite. An example of this will be given, and when this occurs one of the other methods must be used.

3.2 Method I

By (1.4) our approximate system of equations is

$$\mathbf{Ax} = \mathbf{c} \tag{3.1}$$

where $\mathbf{A}$ is a nonsingular coefficient matrix of order n. And by (1.18) the corresponding true system of equations $\mathbf{A^*x^*} = \mathbf{c^*}$ may be written in the form

$$(\mathbf{A} + \delta\mathbf{A})(\mathbf{x} + \delta\mathbf{x}) = \mathbf{c} + \delta\mathbf{c}. \tag{3.2}$$

Thus, corresponding to changes $\delta\mathbf{A}$ and $\delta\mathbf{c}$ the change $\delta\mathbf{x}$ in the solution vector must satisfy (3.2).

Now, on multiplying (3.2) out we have

$$\mathbf{Ax} + \mathbf{A}\delta\mathbf{x} + (\delta\mathbf{A})\mathbf{x} + (\delta\mathbf{A})(\delta\mathbf{x}) = \mathbf{c} + \delta\mathbf{c}.$$

Then, since $\mathbf{Ax} = \mathbf{c}$ it follows that

$$\mathbf{A}\delta\mathbf{x} = \delta\mathbf{c} - (\delta\mathbf{A})\mathbf{x} - (\delta\mathbf{A})(\delta\mathbf{x}). \tag{3.3}$$

Now, premultiplying (3.3) by $\mathbf{B} = \mathbf{A}^{-1}$ we have

$$\delta\mathbf{x} = \mathbf{B}\delta\mathbf{c} - \mathbf{B}(\delta\mathbf{A})\mathbf{x} - \mathbf{B}(\delta\mathbf{A})\delta\mathbf{x}. \tag{3.4}$$

And we call (3.4) our *basic equation of differentials* for the analysis of approximate linear algebraic equations.

Let us now take norms in (3.4). Then

$$\|\delta\mathbf{x}\| = \|\mathbf{B}\delta\mathbf{c} - \mathbf{B}(\delta\mathbf{A})\mathbf{x} - \mathbf{B}(\delta\mathbf{A})\delta\mathbf{x}\|$$
$$\leq \|\mathbf{B}\delta\mathbf{c}\| + \|\mathbf{B}(\delta\mathbf{A})\mathbf{x}\| + \|\mathbf{B}(\delta\mathbf{A})\delta\mathbf{x}\|$$

and hence

$$\|\delta\mathbf{x}\| \leq \|\mathbf{B}\delta\mathbf{c}\| + \|\mathbf{B}\delta\mathbf{A}\|\,\|\mathbf{x}\| + \|\mathbf{B}\delta\mathbf{A}\|\,\|\delta\mathbf{x}\|. \tag{3.5}$$

We now first determine an upper bound for the absolute value of the change that can occur in any one unknown for a change $\delta\mathbf{A}$ in the coefficients and $\delta\mathbf{c}$ in the constants. And then we determine an upper bound for the absolute value of the change that can occur in any one unknown for changes in the coefficients and constants within the limits of their uncertainties.

Since

$$\|\delta\mathbf{x}\|_{\mathrm{I}} = \max_{i}(|\delta\mathbf{x}_i|),$$

we restrict ourselves to first vector and matrix norms, the first matrix norm being compatible with the first vector norm. Then (3.5) becomes

$$\|\delta\mathbf{x}\|_{\mathrm{I}} \leq \|\mathbf{B}\delta\mathbf{c}\|_{\mathrm{I}} + \|\mathbf{B}\delta\mathbf{A}\|_{\mathrm{I}}\,\|\mathbf{x}\|_{\mathrm{I}} + \|\mathbf{B}\delta\mathbf{A}\|_{\mathrm{I}}\,\|\delta\mathbf{x}\|_{\mathrm{I}} \tag{3.6}$$

so that

$$(1 - \|\mathbf{B}\delta\mathbf{A}\|_{\mathrm{I}})\|\delta\mathbf{x}\|_{\mathrm{I}} \leq \|\mathbf{B}\delta\mathbf{c}\|_{\mathrm{I}} + \|\mathbf{B}\delta\mathbf{A}\|_{\mathrm{I}}\,\|\mathbf{x}\|_{\mathrm{I}}.$$

Therefore

$$\|\delta\mathbf{x}\|_{\mathrm{I}} \leq \frac{\|\mathbf{B}\delta\mathbf{c}\|_{\mathrm{I}} + \|\mathbf{B}\delta\mathbf{A}\|_{\mathrm{I}}\,\|\mathbf{x}\|_{\mathrm{I}}}{1 - \|\mathbf{B}\delta\mathbf{A}\|_{\mathrm{I}}} \tag{3.7}$$

provided

$$\|\mathbf{B}\delta\mathbf{A}\|_{\mathrm{I}} < 1. \tag{3.8}$$

The right-hand side of (3.7) is thus an upper bound for the absolute value of the change that can occur in any one unknown for a change $\delta\mathbf{A}$ in the coefficient matrix and $\delta\mathbf{c}$ in the constants.

We now suppose that

$$(|\delta a_{ij}|) = |\delta\mathbf{A}| \leq \Delta\mathbf{A} = (\varepsilon_{ij})$$

and $$(|\delta c_i|) = |\delta\mathbf{c}| \leq \Delta\mathbf{c} = (\varepsilon_i) \tag{3.9}$$

where $\Delta\mathbf{A}$ is the coefficient uncertainty matrix and where $\Delta\mathbf{c} = (\varepsilon_i)$ is the *constants uncertainty vector*. We have thus restricted the changes in the coefficients and constants to lie within the limits of their uncertainties.

Then, since $|\mathbf{B}\delta\mathbf{c}| \leq |\mathbf{B}|\Delta\mathbf{c}$ it clearly follows that

$$\|\mathbf{B}\delta\mathbf{c}\|_{\mathrm{I}} \leq \| \,|\mathbf{B}|\Delta\mathbf{c}\|_{\mathrm{I}}. \tag{3.10}$$

And, similarly, by (1.37) and (2.37)

$$\|\mathbf{B}\delta\mathbf{A}\|_{\mathrm{I}} \leq \| \,|\mathbf{B}|\Delta\mathbf{A}\|_{\mathrm{I}} = \|\mathbf{C}\|_{\mathrm{I}} = \phi_{\mathrm{norm}}. \tag{3.11}$$

Hence, combining (3.7), (3.8), (3.10), and (3.11) we have that

$$\frac{\| \,|\mathbf{B}|\Delta\mathbf{c}\|_{\mathrm{I}} + \phi_{\mathrm{norm}}\,\|\mathbf{x}\|_{\mathrm{I}}}{1 - \phi_{\mathrm{norm}}} \tag{3.12}$$

is an upper bound of the right-hand side of (3.7) subject to (3.9), provided

$$\phi_{\mathrm{norm}} < 1. \tag{3.13}$$

But subject to (3.9) the least upper bounds of the $|\delta x_i|$ are the uncertainties Δx_i. Hence, from (3.7) and (3.12), noting that

$$\|\delta\mathbf{x}\|_{\mathrm{I}} = \max_{i}(|\delta x_i|),$$

we have subject to (3.13) that

$$\Delta x_i \leq \frac{\| \,|\mathbf{B}|\Delta\mathbf{c}\|_{\mathrm{I}} + \phi_{\mathrm{norm}}\,\|\mathbf{x}\|_{\mathrm{I}}}{1 - \phi_{\mathrm{norm}}}, \qquad i = 1, 2, \ldots n. \tag{3.14}$$

Now, our procedure in Method I is to take the right-hand side in (3.14) as our estimates of the uncertainties in the unknowns and to denote these estimates by Δx_i^{I}, the superscript I to indicate estimates by Method I.

Thus:

The estimates of the uncertainties by Method I are given by

$$\Delta x_i^{\mathrm{I}} = \frac{\| \,|\mathbf{B}|\Delta\mathbf{c}\|_{\mathrm{I}} + \phi_{\mathrm{norm}}\,\|\mathbf{x}\|_{\mathrm{I}}}{1 - \phi_{\mathrm{norm}}}, \qquad i = 1, 2, \ldots, n, \tag{3.15}$$

provided $\phi_{\mathrm{norm}} < 1$, and these estimates are upper bounds of the uncertainties Δx_i.

The single value given by Method I is thus an upper bound for the largest uncertainty and no information is available about the variation of the uncertainties in the unknowns. (We may call ϕ_{norm} the *norm ill-conditioning factor* in view of the condition $\phi_{\text{norm}} < 1$ in (3.15).)

And we may note that when $\Delta\mathbf{c} = \mathbf{0}$, i.e., when the right-hand constants are known exactly, then (3.15) becomes:

$$\Delta x_i \leq \Delta x_i^{\text{I}} = \frac{\phi_{\text{norm}}}{1 - \phi_{\text{norm}}} \, \|\mathbf{x}\|_{\text{I}}, \qquad i = 1, 2, \ldots, n,$$

$$\text{when } \Delta\mathbf{c} = \mathbf{0}, \text{ provided } \phi_{\text{norm}} < 1. \tag{3.16}$$

Now, writing

$$\Delta\mathbf{x} = (\Delta x_i), \qquad \Delta\mathbf{x}^{\text{I}} = (\Delta x_i^{\text{I}}) \tag{3.17}$$

we have by (3.16) that:

When $\Delta\mathbf{c} = \mathbf{0}$ and $\phi_{\text{norm}} < 1$ then

$$\frac{\|\Delta\mathbf{x}\|_{\text{I}}}{\|\mathbf{x}\|_{\text{I}}} \leq \frac{\|\Delta\mathbf{x}^{\text{I}}\|_{\text{I}}}{\|\mathbf{x}\|_{\text{I}}} = \phi_{\text{norm}}/(1 - \phi_{\text{norm}}). \tag{3.18}$$

Thus when $\Delta\mathbf{c} = \mathbf{0}$ we may clearly say that $\phi_{\text{norm}}/(1 - \phi_{\text{norm}})$ is an upper bound for the ratio of the uncertainty in the unknown of maximum magnitude to this unknown.

3.3 Special Cases

We deal with two particular cases of (3.15), namely,

$$\varepsilon_{ij} = \varepsilon = \varepsilon_i, \qquad i, j = 1, 2, \ldots, n,$$

and

$$\varepsilon_{ij} = \varepsilon, \qquad \varepsilon_i = 0, \qquad i, j = 1, 2, \ldots, n.$$

Now, when the uncertainties in the coefficients and constants are all equal to ε, i.e.,

$$\varepsilon_{ij} = \varepsilon = \varepsilon_i, \qquad i, j = 1, 2, \ldots, n, \tag{3.19}$$

then

$$\| \, |\mathbf{B}|\Delta\mathbf{c}\|_{\text{I}} = \varepsilon \, \|\mathbf{B}\|_{\text{I}}$$

$$= \frac{1}{n} \, \phi_{\text{norm}} \text{ by (2.40).}$$

Hence, we have by (3.15) that:

When $\varepsilon_{ij} = \varepsilon = \varepsilon_i, \, i, j = 1, 2, \ldots, n$, then

$$\Delta x_i \leq \Delta x_i^{\text{I}} = \frac{\phi_{\text{norm}}}{1 - \phi_{\text{norm}}} \left(\frac{1}{n} + \|\mathbf{x}\|_{\text{I}} \right) \qquad i = 1, 2, \ldots, n \tag{3.20}$$

provided $\phi_{\text{norm}} < 1$ (it being noted that in this case $\phi_{\text{norm}} = \|\mathbf{B}\|_{\text{I}} n\varepsilon$).

And when all the uncertainties in the coefficients are equal to ε but the constants are known exactly, i.e., when

$$\varepsilon_{ij} = \varepsilon, \qquad \varepsilon_i = 0, \qquad i, j = 1, 2, \ldots, n, \tag{3.21}$$

then $\| |\mathbf{B}| \, \Delta\mathbf{c} \|_{\mathrm{I}} = 0$.

Hence, we have by (3.15) that:

When $\varepsilon_{ij} = \varepsilon$, $\varepsilon_i = 0$, $i, j = 1, 2, \ldots, n$, then

$$\Delta x_i \leq \Delta x_i^{\mathrm{I}} = \frac{\phi_{\mathrm{norm}}}{1 - \phi_{\mathrm{norm}}} \|\mathbf{x}\|_{\mathrm{I}}, \qquad i = 1, 2, \ldots, n, \tag{3.22}$$

provided $\phi_{\mathrm{norm}} < 1$ (and $\phi_{\mathrm{norm}} = \|\mathbf{B}\|_{\mathrm{I}} n \varepsilon$ in this case by (2.40)).

Thus, (3.20) and (3.22) give the uncertainties in the unknowns by Method I for the two particular cases being considered, it being emphasized that Method I merely gives an upper bound for the largest uncertainty.

3.4 Example of Determination of Uncertainties by Method I

By applying (3.20) we now determine an upper bound for the uncertainties in the unknowns of the system of equations in (1.43) for the uncertainties given in (1.44), i.e., for an uncertainty of $\varepsilon = 0 \cdot 01$ in each coefficient and constant.

The inverse $\mathbf{B} = \mathbf{A}^{-1}$ of the coefficient matrix $\mathbf{A}$ of the system of equations is given in (1.46). Hence, by (2.40) we have for $\varepsilon = 0 \cdot 01$ that

$$\phi_{\mathrm{norm}} = \|\mathbf{B}\|_{\mathrm{I}} n \varepsilon = (|1 \cdot 6603| + |0 \cdot 8189| + |-1 \cdot 8963|) \times 3 \times 0 \cdot 01$$

i.e.,

$$\phi_{\mathrm{norm}} = 0 \cdot 1313 \tag{3.23}$$

(see (2.48)). Further, by (1.47)

$$\|\mathbf{x}\|_{\mathrm{I}} = \max_i(|x_i|) = 2 \cdot 0438. \tag{3.24}$$

Hence, applying (3.20) we have

$$\Delta x_i \leq \Delta x_i^{\mathrm{I}} = \frac{0 \cdot 1313}{1 - 0 \cdot 1313} (0 \cdot 3333 + 2 \cdot 0438) = 0 \cdot 3593 \tag{3.25}$$

$$i = 1, 2, 3,$$

the condition $\phi_{\mathrm{norm}} < 1$ being clearly satisfied. (Using higher precision throughout the value in (3.25) would actually be $0 \cdot 359172 \ldots$.)

This sets an upper bound of $0 \cdot 3593$ for the uncertainty in each unknown resulting from an uncertainty of $0 \cdot 01$ in each coefficient and right-hand constant. But in our example this bound is far greater than the greatest uncertainty. For it will be seen that upper bounds for the uncertainties in the three unknowns given by Method IV (see (6.71)) are 0·1561, 0·0763, and 0·2400, so that whereas the largest uncertainty is not greater than 0·2400 the method of this chapter gives it as less than or equal to 0·3593.

We may note that had we taken $\varepsilon = 0\cdot1$ in place of $\varepsilon = 0\cdot01$ in (1.44) we would have obtained

$$\phi_{\text{norm}} = 1\cdot313 \quad \text{and} \quad \phi = 0\cdot8611 \tag{3.26}$$

(see (3.23) and (1.48)). Thus, while the system of equations is still not critically ill-conditioned by (2.43) because $\phi = 0\cdot8611 < 1$, nevertheless Method I is inapplicable because $\phi_{\text{norm}} = 1\cdot313 \not< 1$ (see (3.20)). Thus Method I would fail to give finite upper bounds for the uncertainties although the uncertainties are finite. (On the other hand, Method IV, for example, would lead to finite upper bounds for the uncertainties in the unknowns because its condition of application is $\rho(\mathbf{C}) < 1$. And this is satisfied for our example because in our case we have by (2.45) and (3.26) that $\rho(\mathbf{C}) = \phi = 0\cdot8611 < 1$.)

3.5 Uncertainties in Elements of Inverse by Method I

We now apply the theory for determining the uncertainties in the unknowns of an approximate system of equations $\mathbf{Ax} = \mathbf{c}$ to determining the uncertainties in the elements of the inverse $\mathbf{B}$ of the approximate coefficient matrix $\mathbf{A}$.

We have seen in Chapter 1 that the n columns of the inverse $\mathbf{B}$ of a coefficient matrix $\mathbf{A}$ are in fact the n columns of unknowns corresponding to the n unit vectors as right-hand columns of constants. Thus, the elements of the jth column of the inverse $\mathbf{B}$ correspond to the values of the unknowns when the right-hand column of constants is the jth unit vector, i.e.,

$$c_i = 0, \quad i \neq j, \quad c_i = 1, \quad i = j, \quad i = 1, 2, \ldots, n. \tag{3.27}$$

And since the components of the unit vectors are known exactly, i.e.,

$$\Delta \mathbf{c} = \mathbf{0}, \tag{3.28}$$

we use (3.15) in the form it is given in (3.16) to determine the uncertainties in the elements of the inverse.

Now, let us denote the uncertainties in the elements b_{ij} of the inverse by Δb_{ij} and let us denote the estimates of the uncertainties by Method I by

$$\Delta b_{ij}^{\text{I}}, \quad i, j = 1, 2, \ldots, n. \tag{3.29}$$

Further, let us denote the jth column of the inverse $\mathbf{B}$ by $\mathbf{B}_{.j}$, i.e.,

$$\mathbf{B}_{.j} = (b_{1j}, b_{2j}, \ldots, b_{nj})'. \qquad (3.30)$$

Then, to find the uncertainties in the jth column of the inverse we replace Δx_i by Δb_{ij}, Δx_i^{I} by $\Delta b_{ij}^{\mathrm{I}}$, and $\|\mathbf{x}\|_{\mathrm{I}}$ by $\|\mathbf{B}_{.j}\|_{\mathrm{I}}$ in (3.16). Hence:

$$\Delta b_{ij} \leq \Delta b_{ij}^{\mathrm{I}} = \frac{\phi_{\mathrm{norm}}}{1 - \phi_{\mathrm{norm}}} \, \|\mathbf{B}_{.j}\|_{\mathrm{I}}, \qquad i, j = 1, 2, \ldots, n, \qquad (3.31)$$

provided $\phi_{\mathrm{norm}} < 1$.

Thus, (3.31) gives the uncertainties in the elements of the inverse by Method I, an upper bound being given for the uncertainties in each column of the inverse (it being noted that the suffix i does not appear in the expression on the right-hand side of the equality sign in (3.31)).

By way of example, for the coefficient matrix $\mathbf{A}$ in (1.45) we have by (1.46) that

$$\|\mathbf{B}_{.1}\|_{\mathrm{I}} = 1 \cdot 6603, \qquad \|\mathbf{B}_{.2}\|_{\mathrm{I}} = 0 \cdot 8189, \qquad \|\mathbf{B}_{.3}\|_{\mathrm{I}} = 1 \cdot 8963, \quad (3.32)$$

it being recalled that the first vector norm is equal to the absolute value of the element of maximum magnitude in the vector. And for $\varepsilon = 0 \cdot 01$ (see (1.44)) we have by (3.23) that $\phi_{\mathrm{norm}} = 0 \cdot 1313$.

Hence applying (3.31) we have

$$\Delta b_{i1} \leq \Delta b_{i1}^{\mathrm{I}} = \frac{0 \cdot 1313}{1 - 0 \cdot 1313} \times 1 \cdot 6603 = 0 \cdot 2509 \qquad i = 1, 2, 3$$

$$\Delta b_{i2} \leq \Delta b_{i2}^{\mathrm{I}} = \frac{0 \cdot 1313}{1 - 0 \cdot 1313} \times 0 \cdot 8189 = 0 \cdot 1238 \qquad i = 1, 2, 3 \quad (3.33)$$

$$\Delta b_{i3} \leq \Delta b_{i3}^{\mathrm{I}} = \frac{0 \cdot 1313}{1 - 0 \cdot 1313} \times 1 \cdot 8963 = 0 \cdot 2866 \qquad i = 1, 2, 3,$$

the condition of application of (3.31), $\phi_{\mathrm{norm}} < 1$, being clearly satisfied.

And writing

$$\Delta \mathbf{B}^{\mathrm{I}} = (\Delta b_{ij}^{\mathrm{I}}) \qquad (3.34)$$

we have

$$\Delta \mathbf{B}^{\mathrm{I}} = \begin{bmatrix} 0 \cdot 2509 & 0 \cdot 1238 & 0 \cdot 2866 \\ 0 \cdot 2509 & 0 \cdot 1238 & 0 \cdot 2866 \\ 0 \cdot 2509 & 0 \cdot 1238 & 0 \cdot 2866 \end{bmatrix}. \qquad (3.35)$$

This concludes our numerical example.

Now because

$$\|\mathbf{B}_{.j}\|_{\mathrm{I}} = \max_i (|b_{ij}|)$$

it clearly follows that

$$\|\mathbf{B}_{.j}\|_{\mathrm{I}} \leq \max_{k,m} (|b_{km}|), \qquad j = 1, 2, \ldots, n, \qquad (3.36)$$

Hence, corresponding to (3.31) we have the following overall result:

$$\Delta b_{ij} \leq \Delta b_{ij}^{\mathrm{I}} \leq \frac{\phi_{\mathrm{norm}}}{1 - \phi_{\mathrm{norm}}} \max_{k,m}(|b_{km}|), \quad i, j = 1, 2, \ldots, n,$$

provided $\phi_{\mathrm{norm}} < 1$.   (3.37)

And in certain situations this single upper bound for the uncertainties in the elements of the inverse may be good enough.

For our numerical example in this section (3.37) would give

$$\Delta b_{ij} \leq \Delta b_{ij}^{\mathrm{I}} \leq 0\cdot 2866, \qquad i, j = 1, 2, 3 \qquad (3.38)$$

(see (3.33)).

3.6 Conclusion

In concluding our treatment of Method I, we should note that it is clearly necessary first to test the condition of application of the method, i.e., to test whether

$$\|\mathbf{C}\|_{\mathrm{I}} = \phi_{\mathrm{norm}} < 1. \qquad (3.39)$$

And by (2.38) this ensures that the system of equations is not critically ill-conditioned.

But we should also note that the area of application of Method I is more limited than that of Method IV (Chapter 6) which also gives upper bounds. For the condition of application of Method IV

$$\rho(\mathbf{C}) < 1 \qquad (3.40)$$

(see (6.18)) may be satisfied when that for Method I in (3.39) may not be, since $\rho(\mathbf{C}) \leq \|\mathbf{C}\|_{\mathrm{I}}$ (see (2.24)). Further, Method IV gives upper bounds for the individual uncertainties and the largest of these is generally far smaller than the single upper bound given by Method I.

However, Method I is the simpler to apply, and the upper bound of the uncertainties given by Method I may be satisfactory when the solution is required only to a limited accuracy. Method I may also be satisfactory in other situations, for example, when it is required to test whether the unknowns and the elements of the inverse can change sign within the limits of the uncertainties in the coefficients and constants.

CHAPTER 4

Method II: Uncertainties in Unknowns as Linear Functions of Uncertainties in Coefficients and Constants

4.1 Introduction

In this chapter we determine estimates of the uncertainties in the individual unknowns, rather than an upper bound for the maximum uncertainty as in Method I. Although the volume of computation in the two methods is essentially the same, the chief requirement being the inverse of the coefficient matrix, we emphasize that the improved knowledge of the uncertainties obtained by Method II depends on the following assumption being correct. We assume in Method II that the change in value of an unknown is a linear function of the changes in the coefficients and constants and hence we express the uncertainties in the unknowns as linear functions of the uncertainties in the coefficients and constants.

Regarding each unknown as a function of the coefficients and constants, we neglect in effect the partial derivatives of the unknowns with respect to the coefficients and constants of order greater than 1. The method would give the exact value of each uncertainty if the first partial derivatives were to remain constant for variations in the coefficients and constants within the limits of their uncertainties.

But although the estimates by Method II are usually close to the uncertainties, we show that they are in fact lower bounds of the uncertainties subject to the validity of certain assumptions which one might well expect to hold in practice.

It will be seen in Chapter 12 that it is the analysis leading to the estimates by Method II that is used in the statistical approach to approximate equation analysis.

4.2 The Changes in the Unknowns

Now, regarding each unknown as a function of the coefficients and constants, we neglect the partial derivatives of the unknowns with respect to the coefficients and constants of order greater than 1 in determining the changes δx_k $(k = 1, 2, \ldots, n)$ in the solution due to small changes δa_{ij} in the coefficients and δc_i in the constants.

The change δx_k in the unknown x_k due to a small change δa_{11} in a_{11}, say, is then given by the approximate relation

$$\delta x_k \approx \frac{\partial x_k}{\partial a_{11}} \delta a_{11}.$$

And the changes in x_k $(k = 1, 2, \ldots, n)$ for simultaneous changes in all the coefficients and constants are given approximately by

$$\delta x_k = \sum_{i=1}^{n} \sum_{j=1}^{n} \frac{\partial x_k}{\partial a_{ij}} \delta a_{ij} + \sum_{i=1}^{n} \frac{\partial x_k}{\partial c_i} \delta c_i, \qquad k = 1, 2, \ldots, n. \quad (4.1)$$

Expressions for the Partial Derivatives

We now find expressions for the partial derivatives in (4.1) and for simplicity we first consider the system of three equations in three unknowns

$$\begin{aligned}
a_{11}x_1 + a_{12}x_2 + a_{13}x_3 &= c_1 \\
a_{21}x_1 + a_{22}x_2 + a_{23}x_3 &= c_2 \\
a_{31}x_1 + a_{32}x_2 + a_{33}x_3 &= c_3.
\end{aligned} \quad (4.2)$$

Differentiating this system of equations partially with respect to a_{11}, we obtain

$$\begin{aligned}
a_{11}\frac{\partial x_1}{\partial a_{11}} + a_{12}\frac{\partial x_2}{\partial a_{11}} + a_{13}\frac{\partial x_3}{\partial a_{11}} &= -x_1 \\
a_{21}\frac{\partial x_1}{\partial a_{11}} + a_{22}\frac{\partial x_2}{\partial a_{11}} + a_{23}\frac{\partial x_3}{\partial a_{11}} &= 0 \\
a_{31}\frac{\partial x_1}{\partial a_{11}} + a_{32}\frac{\partial x_2}{\partial a_{11}} + a_{33}\frac{\partial x_3}{\partial a_{11}} &= 0.
\end{aligned} \quad (4.3)$$

In matrix notation (4.3) becomes

$$\mathbf{A} \begin{bmatrix} \dfrac{\partial x_1}{\partial a_{11}} \\[2mm] \dfrac{\partial x_2}{\partial a_{11}} \\[2mm] \dfrac{\partial x_3}{\partial a_{11}} \end{bmatrix} = \begin{bmatrix} -x_1 \\[2mm] 0 \\[2mm] 0 \end{bmatrix}. \quad (4.4)$$

Now, premultiplying (4.4) by

$$\mathbf{A}^{-1} \equiv \mathbf{B} = (b_{ij})$$

we have

$$
\begin{bmatrix} \dfrac{\partial x_1}{\partial a_{11}} \\[2mm] \dfrac{\partial x_2}{\partial a_{11}} \\[2mm] \dfrac{\partial x_3}{\partial a_{11}} \end{bmatrix}
= \mathbf{B}
\begin{bmatrix} -x_1 \\ 0 \\ 0 \end{bmatrix}
=
\begin{bmatrix} -x_1 b_{11} \\ -x_1 b_{21} \\ -x_1 b_{31} \end{bmatrix}.
$$

Thus

$$\frac{\partial x_1}{\partial a_{11}} = -x_1 b_{11}, \qquad \frac{\partial x_2}{\partial a_{11}} = -x_1 b_{21}, \qquad \frac{\partial x_3}{\partial a_{11}} = -x_1 b_{31}. \qquad (4.5)$$

The partial derivatives with respect to all the coefficients in the first row of (4.2) are clearly given by

$$\frac{\partial x_k}{\partial a_{1j}} = -x_j b_{k1}, \qquad j, k = 1, 2, 3.$$

More generally, the derivatives with respect to all the coefficients in (4.2) are given by

$$\frac{\partial x_k}{\partial a_{ij}} = -x_j b_{ki}, \qquad i, j, k = 1, 2, 3. \qquad (4.6)$$

Next, we find the partial derivatives with respect to the constants c_i $(i = 1, 2, 3)$. Differentiating (4.2) partially with respect to c_1, we obtain the system of equations

$$a_{11} \frac{\partial x_1}{\partial c_1} + a_{12} \frac{\partial x_2}{\partial c_1} + a_{13} \frac{\partial x_3}{\partial c_1} = 1$$

$$a_{21} \frac{\partial x_1}{\partial c_1} + a_{22} \frac{\partial x_2}{\partial c_1} + a_{23} \frac{\partial x_3}{\partial c_1} = 0 \qquad (4.7)$$

$$a_{31} \frac{\partial x_1}{\partial c_1} + a_{32} \frac{\partial x_2}{\partial c_1} + a_{33} \frac{\partial x_3}{\partial c_1} = 0.$$

Comparing (4.3) and (4.7), it follows by analogy with (4.5) that the partial derivatives with respect to c_1 are

$$\frac{\partial x_k}{\partial c_1} = b_{k1}, \qquad k = 1, 2, 3.$$

More generally, the partial derivatives with respect to all the constants in (4.2) are

$$\frac{\partial x_k}{\partial c_i} = b_{ki} \qquad i, k = 1, 2, 3. \qquad (4.8)$$

Thus, (4.6) and (4.8) give expressions for the partial derivatives required in (4.1) in terms of the unknowns and elements of the inverse of the coefficient matrix for the case $n = 3$. More generally, when dealing with a system of n equations where n is not specified, the partial derivatives in (4.1) are clearly given by

$$\frac{\partial x_k}{\partial a_{ij}} = -x_j b_{ki}, \qquad i, j, k = 1, 2, \ldots, n,$$

and

$$\frac{\partial x_k}{\partial c_i} = b_{ki}, \qquad i, k = 1, 2, \ldots, n.$$

These results can also be derived more formally, and we now give this derivation.

Writing $\mathbf{Ax} = \mathbf{c}$ as $\mathbf{x} = \mathbf{Bc}$ and differentiating partially with respect to a_{ij}, we have

$$\frac{\partial \mathbf{x}}{\partial a_{ij}} = \frac{\partial \mathbf{B}}{\partial a_{ij}} \mathbf{c} = \frac{\partial \mathbf{B}}{\partial a_{ij}} \mathbf{Ax}. \qquad (4.9)$$

Then, differentiating the identity $\mathbf{BA} = \mathbf{I}$ partially with respect to a_{ij}, we have

$$\frac{\partial \mathbf{B}}{\partial a_{ij}} \mathbf{A} = -\mathbf{B} \frac{\partial \mathbf{A}}{\partial a_{ij}}$$

so that (4.9) becomes

$$\frac{\partial \mathbf{x}}{\partial a_{ij}} = -\mathbf{B} \frac{\partial \mathbf{A}}{\partial a_{ij}} \mathbf{x}. \qquad (4.10)$$

Now, the only nonzero element in $\partial \mathbf{A}/\partial a_{ij}$ is in the i,jth position and this element is unity. Hence, on carrying out the multiplications on the right-hand side of (4.10) so that it reduces to a vector, we find on equating corresponding elements in the vectors on either side of the equality sign in (4.10) that

$$\frac{\partial x_k}{\partial a_{ij}} = -x_j b_{ki}, \qquad i, j, k = 1, 2, \ldots, n. \qquad (4.11)$$

And, differentiating $\mathbf{x} = \mathbf{Bc}$ partially with respect to c_i, we have

$$\frac{\partial \mathbf{x}}{\partial c_i} = \mathbf{B} \frac{\partial \mathbf{c}}{\partial c_i}$$

so that

$$\frac{\partial x_k}{\partial c_i} = b_{ki}, \qquad i, k = 1, 2, \ldots, n. \qquad (4.12)$$

The Changes in the Unknowns

Now, substituting the expressions for the partial derivatives in (4.11) and (4.12) into (4.1), we obtain

$$\delta x_k = \sum_{i=1}^{n} \sum_{j=1}^{n} (-x_j b_{ki} \delta a_{ij}) + \sum_{i=1}^{n} b_{ki} \delta c_i, \qquad k = 1, 2, \ldots, n, \qquad (4.13)$$

or, in matrix notation,

$$\delta \mathbf{x} = -\mathbf{B}(\delta \mathbf{A})\mathbf{x} + \mathbf{B}\delta \mathbf{c}. \qquad (4.14)$$

We note that the result in (4.14) is correct to the first order of small quantities because it corresponds to the basic equation of differentials in (3.4) with the last term

$$\mathbf{B}(\delta \mathbf{A})\delta \mathbf{x} \qquad (4.15)$$

neglected, this last term being of the second order of small quantities (assuming that the elements of $\delta \mathbf{A}$ and $\delta \mathbf{x}$ are of the first order of small quantities).

Least Upper Bounds for Changes in Unknowns

We now determine the least upper bounds of the absolute values of the δx_k given by (4.13) for changes in the coefficients and constants within the limits of their uncertainties, i.e., for

$$|\delta a_{ij}| \leq \varepsilon_{ij}, \qquad |\delta c_i| \leq \varepsilon_i, \qquad i, j = 1, 2, \ldots, n. \qquad (4.16)$$

On taking moduli in (4.13), we have

$$|\delta x_k| \leq \sum_{i=1}^{n} \sum_{j=1}^{n} |x_j| \, |b_{ki}| \, |\delta a_{ij}| + \sum_{i=1}^{n} |b_{ki}| \, |\delta c_i|, \qquad k = 1, 2, \ldots, n. \qquad (4.17)$$

Now, when the δa_{ij} and the δc_i are equal in magnitude to the uncertainties in the corresponding coefficients and constants, then the right-hand sides in (4.17) take their greatest values subject to (4.16). Hence, putting

$$|\delta a_{ij}| = \varepsilon_{ij}, \qquad |\delta c_i| = \varepsilon_i, \qquad i, j = 1, 2, \ldots, n,$$

in (4.17), if follows that

$$|\delta x_k| \leq \sum_{i=1}^{n} \sum_{j=1}^{n} |x_j| \, |b_{ki}| \varepsilon_{ij} + \sum_{i=1}^{n} |b_{ki}| \varepsilon_i, \qquad k = 1, 2, \ldots, n, \qquad (4.18)$$

subject to (4.16).

Thus, the right-hand sides in (4.18) are upper bounds of the $|\delta x_k|$ for the changes in the coefficients and constants within the limits of their uncertainties.

And for any given k the equality sign can be reached in (4.18) by choosing the changes δa_{ij} and δc_i in (4.13) equal to the uncertainties and with appropriate signs. For with the changes

$$\delta a_{ij} = \varepsilon_{ij} \operatorname{sign}(-x_j b_{ki}), \qquad \delta c_i = \varepsilon_i \operatorname{sign}(b_{ki}), \qquad i,j = 1, 2, \ldots, n,$$
(4.19)

in (4.13) we obtain

$$
\begin{aligned}
\delta x_k &= \sum_{i=1}^{n} \sum_{j=1}^{n} (-x_j b_{ki}) \varepsilon_{ij} \operatorname{sign}(-x_j b_{ki}) + \sum_{i=1}^{n} b_{ki} \varepsilon_i \operatorname{sign}(b_{ki}) \\
&= \sum_{i=1}^{n} \sum_{j=1}^{n} x_j b_{ki} \varepsilon_{ij} \operatorname{sign}(x_j b_{ki}) + \sum_{i=1}^{n} |b_{ki}| \varepsilon_i \\
&= \sum_{i=1}^{n} \sum_{j=1}^{n} x_j b_{ki} \varepsilon_{ij} \operatorname{sign}(x_j) \operatorname{sign}(b_{ki}) + \sum_{i=1}^{n} |b_{ki}| \varepsilon_i \\
&= \sum_{i=1}^{n} \sum_{j=1}^{n} |x_j| \, |b_{ki}| \varepsilon_{ij} + \sum_{i=1}^{n} |b_{ki}| \varepsilon_i \\
&= \text{right-hand side of (4.18).}
\end{aligned}
$$

And this result holds for $k = 1, 2, \ldots, n$.

Hence, the least upper bounds of the $|\delta x_k|$ given by (4.13) for changes in the coefficients and constants within the limits of their uncertainties are

$$\sum_{i=1}^{n} \sum_{j=1}^{n} |x_j| \, |b_{ki}| \varepsilon_{ij} + \sum_{i=1}^{n} |b_{ki}| \varepsilon_i, \qquad k = 1, 2, \ldots, n. \quad (4.20)$$

4.3   The Uncertainties by Method II

We take the least upper bounds of the absolute values of the δx_k given by (4.13) for the changes in the coefficients and constants within the limits of their uncertainties as the estimates of the uncertainties by Method II. And we denote these estimates by

$$\Delta x_k^{\mathrm{II}}, \qquad k = 1, 2, \ldots, n.$$

Hence by (4.20) we have

$$\Delta x_k^{\mathrm{II}} = \sum_{i=1}^{n} \sum_{j=1}^{n} |x_j| \, |b_{ki}| \varepsilon_{ij} + \sum_{i=1}^{n} |b_{ki}| \varepsilon_i, \qquad k = 1, 2, \ldots, n. \quad (4.21)$$

And by (1.38) we can write (4.21) as

$$\Delta x_k^{\mathrm{II}} = \sum_{j=1}^{n} (c_{kj} |x_j| + |b_{kj}| \varepsilon_j), \qquad k = 1, 2, \ldots, n, \quad (4.22)$$

i.e.,

$$\Delta\mathbf{x}^{\mathrm{II}} = \mathbf{C}|\mathbf{x}| + |\mathbf{B}|\Delta\mathbf{c} \qquad (4.23)$$

where

$$\Delta\mathbf{x}^{\mathrm{II}} = (\Delta x_k^{\mathrm{II}}), \qquad |\mathbf{x}| = (|x_j|), \qquad \Delta\mathbf{c} = (\varepsilon_j).$$

Now, we have seen by (4.19) and (4.20) that for a given k we must choose the following changes

$$\delta a_{ij} = \varepsilon_{ij}\,\mathrm{sign}(-x_j b_{ki}), \qquad \delta c_i = \varepsilon_i\,\mathrm{sign}(b_{ki}), \qquad i, j = 1, 2, \ldots, n \qquad (4.24)$$

for a positive change δx_k in (4.13) as large as Δx_k^{II} in (4.21), i.e., for

$$\delta x_k = \Delta x_k^{\mathrm{II}}.$$

And let us note that if we choose changes δa_{ij} and δc_i in (4.13) of opposite sign to that in (4.24), i.e., if we choose

$$\delta a_{ij} = \varepsilon_{ij}\,\mathrm{sign}(x_j b_{ki}), \qquad \delta c_i = -\varepsilon_i\,\mathrm{sign}(b_{ki}), \qquad i, j = 1, 2, \ldots, n, \qquad (4.25)$$

then

$$\delta x_k = -\Delta x_k^{\mathrm{II}}.$$

Finally, we may note that Method II gives the true values of the uncertainties when there are uncertainties in the constants only but not in the coefficients. For the neglected term in (4.15) would then in fact be zero, i.e., $\mathbf{B}(\delta\mathbf{A})\,\delta\mathbf{x} = \mathbf{0}$. Thus, by (4.21) we have

$$\Delta x_k = \Delta x_k^{\mathrm{II}} = \sum_{i=1}^{n} |b_{ki}|\varepsilon_i \qquad \text{for any } \Delta\mathbf{c} = (\varepsilon_i) \text{ when } \Delta\mathbf{A} = (\varepsilon_{ij}) = \mathbf{0},$$

$$k = 1, 2, \ldots, n, \qquad (4.26)$$

i.e.,

$$\Delta\mathbf{x} = \Delta\mathbf{x}^{\mathrm{II}} = |\mathbf{B}|\Delta\mathbf{c} \text{ for any } \Delta\mathbf{c} \text{ when } \Delta\mathbf{A} = \mathbf{0},$$

where

$$\Delta\mathbf{x} = (\Delta x_k).$$

4.4 Special Cases

For the case where the uncertainty in each coefficient and constant is ε, i.e.,

$$\varepsilon_{ij} = \varepsilon = \varepsilon_i, \qquad i, j = 1, 2, \ldots, n,$$

we have by (4.21) that

$$\Delta x_k^{\mathrm{II}} = \left(\sum_{i=1}^{n}\sum_{j=1}^{n} |x_j|\,|b_{ki}| + \sum_{i=1}^{n} |b_{ki}|\right)\varepsilon$$

$$= \left(1 + \sum_{j=1}^{n} |x_j|\right)\left(\sum_{i=1}^{n} |b_{ki}|\right)\varepsilon, \qquad k = 1, 2, \ldots, n.$$

Thus:

$$\Delta x_k^{II} = (1 + x)b_k\varepsilon, \qquad k = 1, 2, \ldots, n,$$
$$\text{when } \varepsilon_{ij} = \varepsilon = \varepsilon_i, \qquad i, j = 1, 2, \ldots, n \qquad (4.27)$$

where

$$x \equiv \sum_{j=1}^{n} |x_j| = \begin{array}{l}\text{sum of absolute values of unknowns of} \\ \text{approximate system of equations } \mathbf{Ax} = \mathbf{c}\end{array} \quad (4.28)$$

and where b_k is defined in (1.41).

Corresponding to (4.27), the relevant result when there are uncertainties in the coefficients only (but not in the constants) is:

$$\Delta x_k^{II} = xb_k\varepsilon, \qquad k = 1, 2, \ldots, n$$
$$\text{when } \varepsilon_{ij} = \varepsilon, \qquad i, j = 1, 2, \ldots, n, \qquad \text{and } \Delta\mathbf{c} = \mathbf{0}. \qquad (4.29)$$

And when only the constants have uncertainties (but not the coefficients) then corresponding to (4.27) we have by (4.26) that:

$$\Delta x_k = \Delta x_k^{II} = \varepsilon b_k, \qquad k = 1, 2, \ldots, n$$
$$\text{when } \Delta\mathbf{A} = \mathbf{0} \quad \text{and} \quad \varepsilon_i = \varepsilon, \qquad i = 1, 2, \ldots, n. \qquad (4.30)$$

Let us now write

$$\mathbf{b} = (b_k) \qquad (4.31)$$

and call $\mathbf{b}$ the *row sum vector of the inverse* $\mathbf{B}$, it being recalled that

$$b_k = \sum_{j=1}^{n} (|b_{kj}|), \qquad k = 1, 2, \ldots, n.$$

Then we may state the results in (4.27), (4.29), and (4.30) more briefly as

$$\Delta\mathbf{x}^{II} = (1 + x)\varepsilon\mathbf{b} \qquad \text{when } \varepsilon_{ij} = \varepsilon = \varepsilon_i, i, j = 1, 2, \ldots, n \qquad (4.32)$$

$$\Delta\mathbf{x}^{II} = x\varepsilon\mathbf{b} \qquad \text{when } \Delta\mathbf{c} = \mathbf{0} \text{ and } \varepsilon_{ij} = \varepsilon, i, j = 1, 2, \ldots, n \qquad (4.33)$$

$$\Delta\mathbf{x} = \Delta\mathbf{x}^{II} = \varepsilon\mathbf{b} \qquad \text{when } \Delta\mathbf{A} = \mathbf{0} \text{ and } \varepsilon_i = \varepsilon, i = 1, 2, \ldots, n. \qquad (4.34)$$

4.5   Examples of Determination of Uncertainties by Method II

Example 1

We now determine the uncertainties by Method II for our numerical example in (1.43) and (1.44).

By (4.27) we have

$$\Delta x_1^{II} = (1 + x)b_1\varepsilon = (1 + 4\cdot0130) \times 2\cdot8451 \times 0\cdot01 = 0\cdot1426$$

since by (1.47) and (4.28)

$$x = |0 \cdot 9730| + |-0 \cdot 9962| + |2 \cdot 0438| = 4 \cdot 0130$$

and by (1.46) and (1.41)

$$b_1 = |-1 \cdot 0574| + |0 \cdot 0724| + |1 \cdot 7153| = 2 \cdot 8451.$$

Proceeding similarly for Δx_2^{II} and Δx_3^{II}, Method II gives the following estimates of the uncertainties for x_1, x_2, and x_3 for the system of equations in (1.43) for $\varepsilon = 0 \cdot 01$:

$$\Delta x_1^{II} = 0 \cdot 1426, \qquad \Delta x_2^{II} = 0 \cdot 0697, \qquad \Delta x_3^{II} = 0 \cdot 2193. \qquad (4.35)$$

We note that the upper bound for the uncertainties by Method I for this example is $0 \cdot 3593$ (see (3.25)) whereas by (4.35) the maximum estimate by Method II is $0 \cdot 2193$.

Example 2

It is interesting to note how the actual changes in the unknowns compare with the uncertainties in a particular example.

Let us then suppose that the following system of equations

$$\begin{aligned} 0 \cdot 526x_1 + 0 \cdot 861x_2 + 0 \cdot 485x_3 &= 0 \cdot 635 \\ 0 \cdot 942x_1 - 0 \cdot 467x_2 + 0 \cdot 854x_3 &= 3 \cdot 117 \\ 0 \cdot 874x_1 + 0 \cdot 552x_2 + 0 \cdot 256x_3 &= 0 \cdot 834 \end{aligned} \qquad (4.36)$$

is the true system of equations corresponding to (1.43).

But noting that on rounding the system of equations in (4.36) to two decimal places we obtain (1.43), let us assume that $\varepsilon = 0 \cdot 005$ in (1.44) instead of $0 \cdot 01$. Then, since by (4.27) the Δx_k^{II} are proportional to ε the uncertainties for $\varepsilon = 0 \cdot 005$ are half those in (4.35). Thus

$$\Delta x_1^{II} = 0 \cdot 0713, \qquad \Delta x_2^{II} = 0 \cdot 0348, \qquad \Delta x_3^{II} = 0 \cdot 1096 \qquad (4.37)$$

for (1.43) when $\varepsilon = 0 \cdot 005$ instead of $0 \cdot 01$ in (1.44).

Now, it is easy to check by direct substitution that the solution of (4.36) is in fact

$$x_1 = 1, \qquad x_2 = -1, \qquad x_3 = 2. \qquad (4.38)$$

Hence comparing the solutions of (1.43) and (4.36) we have by (1.47) and (4.38) that

$$\begin{aligned} |\delta x_1| &= |1 - 0 \cdot 9730| = 0 \cdot 0270 \\ |\delta x_2| &= |-1 + 0 \cdot 9962| = 0 \cdot 0038 \\ |\delta x_3| &= |2 - 2 \cdot 0438| = 0 \cdot 0438. \end{aligned} \qquad (4.39)$$

Clearly, the errors or changes in (4.39) fall well within the estimates of the uncertainties in (4.37).

4.6 A Few General Remarks About Uncertainties

To obtain the estimate of the uncertainty by Method II in any one unknown, we in effect (see (4.24)) make changes of maximum magnitude in the coefficients and constants (i.e., equal to their uncertainties in magnitude) and with appropriate signs so as to produce a change of maximum magnitude in the unknown as given by the approximate relation (4.13). Let us suppose for the moment that the Δx_k^{II} are in fact close to the Δx_k. Then the approximate system of equations could differ from the true system in such a way, i.e., the errors in the coefficients and constants could be such that the error in an unknown is as large as its uncertainty. But the actual errors in the coefficients and constants are seldom such as to produce this effect.

Further, if a set of errors in the coefficients and constants produces an error in an unknown as large in magnitude as its uncertainty, then it will not necessarily produce errors in the other unknowns equal in magnitude to their uncertainties.

Hence, the actual errors in the unknowns usually fall well within the limits given by the computed uncertainties.

4.7 Uncertainties in Elements of Inverse by Method II

Let us denote the estimate of the uncertainty by Method II in the element b_{km} by $\Delta b_{km}^{\text{II}}$. And let us recall that the mth column of the inverse corresponds to the solution vector when the right-hand column of constants is the mth unit vector, i.e., the column vector with its mth element unity and all its other elements zero.

Hence, to obtain the estimates of the uncertainties by Method II in the elements in the mth column of the inverse we clearly put

$$\Delta x_k^{\text{II}} = \Delta b_{km}^{\text{II}}, \qquad |x_j| = |b_{jm}|, \qquad \varepsilon_i = 0$$

in (4.21), the ε_i being zero because the unit vectors leading to the elements of the inverse are known exactly.

This gives us

$$\Delta b_{km}^{\text{II}} = \sum_{i=1}^{n} \sum_{j=1}^{n} |b_{ki}|\, \varepsilon_{ij}\, |b_{jm}|, \qquad k, m = 1, 2, \ldots, n, \qquad (4.40)$$

i.e.,

$$\Delta \mathbf{B}^{\text{II}} = |\mathbf{B}|\, \Delta \mathbf{A}\, |\mathbf{B}| = \mathbf{C}|\mathbf{B}| \qquad (4.41)$$

where
$$\Delta \mathbf{B}^{\text{II}} = (\Delta b_{km}^{\text{II}}).$$

Now let us consider the case when the uncertainties in the coefficients are all equal to ε, i.e.,

$$\varepsilon_{ij} = \varepsilon, \qquad i, j = 1, 2, \ldots, n.$$

Then, to obtain the result corresponding to (4.29) for the uncertainties in the elements in the mth column of the inverse we replace x in (4.29) by the sum of the absolute values of the elements in the mth column of the inverse (see (4.28)). Thus

$$\Delta b_{km}^{II} = b_k b_{\cdot m} \varepsilon, \qquad k, m = 1, 2, \ldots, n$$

when

$$\varepsilon_{ij} = \varepsilon, \qquad i, j = 1, 2, \ldots, n \tag{4.42}$$

where

$$b_{\cdot m} = \sum_{j=1}^{n} |b_{jm}| = \frac{\text{sum of absolute values of elements in } m\text{th}}{\text{column of inverse}} \tag{4.43}$$

and where b_k is defined in (1.41).

And it clearly follows from (4.42) that

$$\Delta b_{km}^{II} \leq (\|\mathbf{B}\|_I \|\mathbf{B}\|_{II})\varepsilon, \qquad k, m = 1, 2, \ldots, n \tag{4.44}$$

when all the $\varepsilon_{ij} = \varepsilon$, it being recalled that the first matrix norm is equal to the maximum row sum (i.e., $\|\mathbf{B}\|_I = \max_k(b_k)$), while the second matrix norm represents the maximum column sum. Thus (4.44) gives an upper bound for the maximum of the estimates of the uncertainties in the elements of the inverse by Method II when all the $\varepsilon_{ij} = \varepsilon$, the right-hand side in (4.44) being the largest of the right-hand sides in (4.42), i.e.,

$$(\|\mathbf{B}\|_I \|\mathbf{B}\|_{II})\varepsilon = \max_{k,m}(b_k b_{\cdot m} \varepsilon). \tag{4.45}$$

By way of a numerical example, let us determine the uncertainties in the elements of the inverse in (1.46) of the coefficient matrix in (1.45) for an uncertainty of 0·01 in each coefficient.

From (1.46) we have

$$\begin{aligned} b_1 &= |-1{\cdot}0574| + |0{\cdot}0724| + |1{\cdot}7153| = 2{\cdot}8451 \\ b_2 &= |0{\cdot}8878| + |-0{\cdot}5017| + |0{\cdot}0013| = 1{\cdot}3908 \\ b_3 &= |1{\cdot}6603| + |0{\cdot}8189| + |1{\cdot}8963| = 4{\cdot}3755 \end{aligned} \tag{4.46}$$

and

$$\begin{aligned} b_{\cdot 1} &= |-1{\cdot}0574| + |0{\cdot}8878| + |1{\cdot}6603| = 3{\cdot}6055 \\ b_{\cdot 2} &= |0{\cdot}0724| + |-0{\cdot}5017| + |0{\cdot}8189| = 1{\cdot}3930 \\ b_{\cdot 3} &= |1{\cdot}7153| + |0{\cdot}0013| + |-1{\cdot}8963| = 3{\cdot}6129. \end{aligned} \tag{4.47}$$

Hence for $\varepsilon = 0{\cdot}01$ we have by (4.42) that

$$\Delta \mathbf{B}^{II} = (\Delta b_{km}^{II}) = \begin{bmatrix} 0{\cdot}1026 & 0{\cdot}0396 & 0{\cdot}1028 \\ 0{\cdot}0501 & 0{\cdot}0194 & 0{\cdot}0502 \\ 0{\cdot}1578 & 0{\cdot}0610 & 0{\cdot}1581 \end{bmatrix}. \tag{4.48}$$

Comparing (4.48) and (3.35), we note that the estimates of the uncertainties in the elements of the inverse by Method II are considerably less than the upper bounds in (3.35).

And

$$\|\mathbf{B}\|_{\mathrm{I}} = 4{\cdot}3755 \qquad \text{and} \qquad \|\mathbf{B}\|_{\mathrm{II}} = 3{\cdot}6129$$

from (4.46) and (4.47) so that (4.44) gives

$$\Delta b_{km}^{\mathrm{II}} \le \|\mathbf{B}\|_{\mathrm{I}} \|\mathbf{B}\|_{\mathrm{II}} \varepsilon = 4{\cdot}3755 \times 3{\cdot}6129 \times 0{\cdot}01 = 0{\cdot}1581,$$
$$k, m = 1, 2, 3. \tag{4.49}$$

The right-hand side in (4.49) is of course the largest element in (4.48), and this is less than the corresponding value of $0{\cdot}2866$ given by Method I (see (3.38) and (3.35)).

4.8   The Δx_k^{II} as Lower Bounds of the Δx_k

We now proceed to show that the Δx_k^{II} are in fact lower bounds of the uncertainties in the unknowns subject to certain assumptions which one might well expect will hold in practice when the uncertainties involved are sufficiently small.

We begin by considering the basic equation of differentials (3.4)

$$\delta x_m = -\sum_{i=1}^{n}\sum_{j=1}^{n} b_{mi}\, \delta a_{ij} x_j + \sum_{i=1}^{n} b_{mi}\, \delta c_i - \sum_{i=1}^{n}\sum_{j=1}^{n} b_{mi}\, \delta a_{ij}\, \delta x_j,$$
$$m = 1, 2, \ldots, n \tag{4.50}$$

with the changes δa_{ij} and δc_i in (4.24) for some fixed k, i.e., with

$$\delta a_{ij} = \varepsilon_{ij}\, \mathrm{sign}(-x_j b_{ki}), \qquad \delta c_i = \varepsilon_i\, \mathrm{sign}(b_{ki}), \qquad i, j = 1, 2, \ldots, n. \tag{4.51}$$

And let us denote the resulting set of changes δx_m by

$$d_{mk} = \delta x_m, \qquad m = 1, 2, \ldots, n, \tag{4.52}$$

the second suffix k in d_{mk} corresponding to the k in (4.51).

Then, by (4.50), (4.51), and (4.52), we have

$$d_{mk} = \sum_{i=1}^{n}\sum_{j=1}^{n} \varepsilon_{ij}|x_j| b_{ki}\, \mathrm{sign}(b_{ki}) + \sum_{i=1}^{n} \varepsilon_i b_{mi}\, \mathrm{sign}(b_{ki})$$
$$+ \sum_{i=1}^{n}\sum_{j=1}^{n} \varepsilon_{ij} d_{jk} b_{mi}\, \mathrm{sign}(x_j b_{ki}), \qquad m = 1, 2, \ldots, n. \tag{4.53}$$

And now consider the basic equation of differentials with changes δa_{ij} and δc_i as in (4.25), i.e.,

$$\delta a_{ij} = \varepsilon_{ij} \operatorname{sign}(x_j b_{ki}), \qquad \delta c_i = -\varepsilon_i \operatorname{sign}(b_{ki}), \qquad i, j = 1, 2, \ldots, n,$$
(4.54)

each change being of opposite sign to that in (4.51). Then, denoting the changes δx_m in the unknowns by

$$e_{mk} = \delta x_m, \qquad m = 1, 2, \ldots, n$$

we have from (4.50) that

$$e_{mk} = -\sum_{i=1}^{n} \sum_{j=1}^{n} \varepsilon_{ij} |x_i| b_{mi} \operatorname{sign}(b_{ki}) - \sum_{i=1}^{n} \varepsilon_i b_{mi} \operatorname{sign}(b_{ki})$$
$$- \sum_{i=1}^{n} \sum_{j=1}^{n} \varepsilon_{ij} e_{jk} b_{mi} \operatorname{sign}(x_j b_{ki}), \qquad m = 1, 2, \ldots, n.$$
(4.55)

Let us now suppose that the effect of the second-order term (i.e., the last term) on the right-hand side in (4.53) and in (4.55) is negligible in contributing to the value of the corresponding left-hand side. Then we may say that while the d_{mk} ($m = 1, 2, \ldots, n$) in (4.53) are of opposite sign to the e_{mk} ($m = 1, 2, \ldots, n$) in (4.55), they nevertheless do not differ much in magnitude, i.e.,

$$d_{mk} \approx -e_{mk}, \qquad m = 1, 2, \ldots, n.$$
(4.56)

Now, putting $m = k$ in (4.53) we have

$$d_{kk} = \sum_{i=1}^{n} \sum_{j=1}^{n} \varepsilon_{ij} |x_j| \, |b_{ki}| + \sum_{i=1}^{n} \varepsilon_i |b_{ki}|$$
$$+ \sum_{i=1}^{n} \sum_{j=1}^{n} |b_{ki}| \varepsilon_{ij} d_{jk} \operatorname{sign}(x_j).$$
(4.57)

Hence, by (4.21) and (1.38) we have

$$d_{kk} = \Delta x_k^{\mathrm{II}} + \sum_{j=1}^{n} c_{kj} d_{jk} \operatorname{sign}(x_j),$$
(4.58)

the second term on the right-hand side being of the second order of small quantities, assuming that the c_{kj} and d_{jk} are of the first order of small quantities. (The c_{kj} may be regarded as being of the first order of small quantities if the ε_{ij} are of the first order of small quantities and the elements of the inverse are not too large.)

Similarly, putting $m = k$ in (4.55) we have

$$e_{kk} = -\Delta x_k^{II} - \sum_{j=1}^{n} c_{kj}e_{jk}\, \text{sign}(x_j). \qquad (4.59)$$

And in view of (4.56) we expect the second-order terms in (4.58) and (4.59) to be nearly equal in magnitude and of opposite sign, i.e., we expect that

$$\sum_{j=1}^{n} c_{kj}d_{jk}\, \text{sign}(x_j) \approx - \sum_{j=1}^{n} c_{kj}e_{jk}\, \text{sign}(x_j). \qquad (4.60)$$

Now, to show that Δx_k^{II} is a lower bound of the uncertainty Δx_k all we require is the assumption that

$$\sum_{j=1}^{n} c_{kj}d_{jk}\, \text{sign}(x_j) \qquad \text{and} \qquad \sum_{j=1}^{n} c_{kj}e_{jk}\, \text{sign}(x_j) \qquad (4.61)$$

are of opposite sign, an assumption that will in practice be satisfied in view of (4.60).

Then, according as the last term in (4.58) is positive or negative we have from (4.58) and (4.59) that

$$d_{kk} \geq \Delta x_k^{II} \qquad \text{or} \qquad e_{kk} \leq -\Delta x_k^{II}, \qquad (4.62)$$

respectively, the equality signs in (4.62) allowing for the case where the second-order terms in (4.61) are both zero. Thus

$$|d_{kk}| \geq \Delta x_k^{II} \qquad \text{or} \qquad |e_{kk}| \geq \Delta x_k^{II}. \qquad (4.63)$$

But d_{kk} and e_{kk} are changes that can occur in x_k within the limits of the uncertainties in the coefficients and constants, the changes in the coefficients and constants being considered being those in (4.51) and (4.54). Hence,

$$\Delta x_k \geq |d_{kk}| \qquad \text{and} \qquad \Delta x_k \geq |e_{kk}|. \qquad (4.64)$$

Hence, by (4.63) and (4.64)

$$\Delta x_k^{II} \leq \Delta x_k. \qquad (4.65)$$

And (4.65) must of course hold for $k = 1, 2, \ldots, n$ if the expressions in (4.61) are of opposite sign for each of $k = 1, 2, \ldots, n$.

Thus,

$$\Delta x_k^{II} \leq \Delta x_k, \qquad k = 1, 2, \ldots, n, \qquad (4.66)$$

i.e., the estimates of the uncertainties by Method II are lower bounds of the true uncertainties, subject to the assumption that the expressions in (4.61) are of opposite sign for each of $k = 1, 2, \ldots, n$.

Further, assuming that (4.60) holds for each of $k = 1, 2, \ldots, n$, we have the following approximate result from (4.58) and (4.59):

$$\Delta x_k^{II} \approx \tfrac{1}{2}(d_{kk} - e_{kk}), \qquad k = 1, 2, \ldots, n. \tag{4.67}$$

And this result could for example be used to estimate e_{kk} from d_{kk} and Δx_k^{II}.

Let us now regard the d_{mk} and e_{mk} in (4.53) and (4.55) as the elements of matrices **D** and **E**, respectively, i.e., we write

$$\mathbf{D} = (d_{mk}) \qquad \text{and} \qquad \mathbf{E} = (e_{mk}). \tag{4.68}$$

Then, for our numerical example in (1.43) and (1.44) we obtain

$$\mathbf{D} = \begin{bmatrix} \underline{0\cdot1448} & 0\cdot0288 & -0\cdot1371 \\ -0\cdot0706 & \underline{0\cdot0684} & 0\cdot0195 \\ -0\cdot1393 & -0\cdot0519 & \underline{0\cdot2222} \end{bmatrix},$$

$$\mathbf{E} = \begin{bmatrix} \underline{-0\cdot1405} & -0\cdot0299 & 0\cdot1336 \\ 0\cdot0686 & \underline{-0\cdot0710} & -0\cdot0190 \\ 0\cdot1352 & 0\cdot0539 & \underline{-0\cdot2165} \end{bmatrix},$$

$$\tag{4.69}$$

the diagonal elements in the two matrices being underlined. And we emphasize that to obtain each column of **D** and **E** above requires the solution of a system of equations of order 3, the computational procedures to be adopted being discussed in the next chapter.

Now, first comparing corresponding elements in **D** and **E** above we see that (4.56) holds, i.e., $d_{mk} \approx -e_{mk}$, $m, k = 1, 2, 3$.

Secondly, it is clear that in our example the Δx_k^{II} are lower bounds of the uncertainties because by (4.35), (4.69), and (4.64) we have

$$\begin{aligned} \Delta x_1^{II} &= 0\cdot1426 < 0\cdot1448 = |d_{11}| \le \Delta x_1 \\ \Delta x_2^{II} &= 0\cdot0697 < 0\cdot0710 = |e_{22}| \le \Delta x_2 \\ \Delta x_3^{II} &= 0\cdot2193 < 0\cdot2222 = |d_{33}| \le \Delta x_3. \end{aligned} \tag{4.70}$$

And thirdly it is clear that (4.67) holds very nearly in our example. For

$$\begin{aligned} \Delta x_1^{II} &= 0\cdot1426 \quad \text{while} \quad \tfrac{1}{2}(d_{11} - e_{11}) = \tfrac{1}{2}(0\cdot1448 + 0\cdot1405) = 0\cdot1426, \\ \Delta x_2^{II} &= 0\cdot0697 \quad \text{while} \quad \tfrac{1}{2}(d_{22} - e_{22}) = \tfrac{1}{2}(0\cdot0684 + 0\cdot0710) = 0\cdot0697, \\ \Delta x_3^{II} &= 0\cdot2193 \quad \text{while} \quad \tfrac{1}{2}(d_{33} - e_{33}) = \tfrac{1}{2}(0\cdot2222 + 0\cdot2165) = 0\cdot2193. \end{aligned}$$
$$\tag{4.71}$$

The closeness of the results in (4.71) indicates that (4.60) is a good approximation for our example and it follows that the assumption that the expressions in (4.61) are of opposite sign is certainly valid for our example. Hence, as checked in (4.70), the Δx_k^{II} must be lower bounds of the Δx_k for our example. And, as indicated earlier, we expect this result to hold in all but exceptional cases.

In the next method, Method III, the estimates of the uncertainties are always lower bounds, the $\Delta x_k^{\mathrm{III}}$ being expected to be closer to the uncertainties than the Δx_k^{II}. And, subject to certain assumptions being valid, the $\Delta x_k^{\mathrm{III}}$ are in fact equal to the uncertainties, each $\Delta x_k^{\mathrm{III}}$ being the larger of the corresponding $|d_{kk}|$ and $|e_{kk}|$ in (4.58) and (4.59).

CHAPTER 5

Method III: Uncertainties in Unknowns Assuming Partial Derivatives do not Change Sign

5.1 Introduction

In Method III we assume that the first partial derivatives of the unknowns with respect to the coefficients and constants do not change sign within the limits of the uncertainties in the coefficients and constants. Thus:

In Method III we assume that the

$$\frac{\partial x_k^*}{\partial a_{ij}^*} \quad \text{and} \quad \frac{\partial x_k^*}{\partial c_i^*}, \quad i, j, k = 1, 2, \ldots, n, \tag{5.1}$$

do not change sign for the a_{ij}^* and c_i^* given in (1.7).

This is a lesser assumption than that made in Method II where in effect we assumed that the derivatives in (5.1) remain constant within the limits of the uncertainties, the assumption in Method II being equivalent to neglecting the higher order derivatives.

In Method III we determine both the maximum possible increase and the maximum possible decrease in each unknown for changes in the coefficients and constants within the limits of the uncertainties. In this way we obtain the intervals of uncertainty, and we take the larger of each increase and corresponding decrease as the estimate of the uncertainty in the unknown by Method III.

Method III leads to the true intervals of uncertainty when the assumptions made in it are valid and hence when certain derived conditions of application of the method are satisfied.

But the same results can be obtained by linear programming techniques (Chapter 11). And further, linear programming gives the true intervals of uncertainty when Method III fails to do so.

Method III, however, is the more direct procedure so it is to be preferred when applicable.

And it may be pointed out that when the assumption in (5.1) cannot be shown to be valid it is nevertheless possible for Method III to be combined with Method IV of Chapter 6 to obtain intervals containing the true intervals of uncertainty (Chapter 8). Under the same circumstances interval arithmetic techniques can also be combined with Method III to obtain intervals which contain the true intervals of uncertainty (Section 9.8).

5.2 Method III

To determine the maximum possible increase in the unknown x_k for changes in the coefficients and constants within the limits of the uncertainties, we introduce changes in the coefficients and constants all equal to the uncertainties and each with sign so as to increase the value of x_k. Thus, to determine the maximum possible increase in x_k we put

$$\delta a_{ij} = \varepsilon_{ij}\,\text{sign}\!\left(\frac{\partial x_k}{\partial a_{ij}}\right), \qquad \delta c_i = \varepsilon_i\,\text{sign}\!\left(\frac{\partial x_k}{\partial c_i}\right), \qquad i, i = 1, 2, \ldots, n,$$

(5.2)

where the derivatives are taken at $\mathbf{x}$ where $\mathbf{Ax} = \mathbf{c}$ (see (1.4)). And by (4.11) and (4.12) this becomes

$$\delta a_{ij} = \varepsilon_{ij}\,\text{sign}(-x_j b_{ki}), \qquad \delta c_i = \varepsilon_i\,\text{sign}(b_{ki}), \qquad i, j = 1, 2, \ldots, n.$$

(5.3)

With these changes we obtain the system of equations

$$\sum_{j=1}^{n}(a_{ij} + \varepsilon_{ij}\,\text{sign}(-x_j b_{ki}))x_j^* = c_i + \varepsilon_i\,\text{sign}(b_{ki}), \qquad i = 1, 2, \ldots, n,$$

(5.4)

where the x_j^* refer to the perturbed system of equations.

The maximum increase in x_k is then given by

$$x_k^* - x_k,$$

(5.5)

i.e., by the kth component of

$$\delta \mathbf{x} = \mathbf{x}^* - \mathbf{x}.$$

(5.6)

Now, we may note that the changes in (5.3) are the same as those in (4.51) so that $\delta \mathbf{x} = \mathbf{x}^* - \mathbf{x}$ in (5.6) has components $d_{1k}, d_{2k}, \ldots, d_{nk}$ (see (4.52)), i.e.,

$$\delta \mathbf{x} = \mathbf{x}^* - \mathbf{x} = (d_{1k}, d_{2k}, \ldots, d_{nk})'.$$

For, theoretically, though not necessarily computationally, it is immaterial whether we determine the increment vector $\delta\mathbf{x}$ as the difference between two solution vectors or by using the basic equation of differentials. Thus, d_{kk} is equal to the maximum increase in x_k given in (5.5), i.e., $d_{kk} = x_k^* - x_k$.

And in general we have:

> The diagonal elements of the matrix $\mathbf{D}$ (see (4.68)) give the maximum possible increases in x_k ($k = 1, 2, \ldots, n$),   (5.7) subject to the assumption in Method III being valid.

Then, to determine the maximum decrease in x_k within the limits of the uncertainties, we introduce changes in the coefficients and constants as given in (5.3) but with opposite sign, i.e., we put

$$\delta a_{ij} = \varepsilon_{ij} \, \mathrm{sign}(x_j b_{ki}), \qquad \delta c_i = -\varepsilon_i \, \mathrm{sign}(b_{ki}), \qquad i, j = 1, 2, \ldots, n. \tag{5.8}$$

Solving the perturbed system of equations with these changes, we obtain the maximum possible decrease in x_k for changes in the coefficients and constants within the limits of their uncertainties.

Now, noting that the changes in (5.8) are the same as those in (4.54), we see that it follows that the maximum decrease in x_k is equal to e_{kk} given by (4.55). Thus, corresponding to (5.7) we have:

> The diagonal elements of the matrix $\mathbf{E}$ (see (4.68)) give the maximum possible decreases in x_k ($k = 1, 2, \ldots, n$)   (5.9) subject to the assumption in Method III being valid.

But, subject to the assumption in Method III in (5.1), the changes in x_k ($k = 1, 2, \ldots, n$) cannot be negative for the δa_{ij} and δc_i in (5.3). Hence, subject to the assumption in (5.1), it follows by (5.7) that the diagonal elements of $\mathbf{D}$ cannot be negative. Similarly, it follows by (5.8) and (5.9) that the diagonal elements of $\mathbf{E}$ cannot be positive subject to the assumption in (5.1). (For a numerical example see (4.69).)

Hence, it follows by (5.7) and (5.9) that:

> The intervals of uncertainty in the unknowns are given by
> $$[x_k + e_{kk}, \ x_k + d_{kk}], \qquad k = 1, 2, \ldots, n, \tag{5.10}$$
> subject to the assumption in Method III being valid.

And we define the uncertainties in the unknowns given by Method III by

$$\Delta x_k^{\mathrm{III}} = \max(|e_{kk}|, |d_{kk}|), \qquad k = 1, 2, \ldots, n. \tag{5.11}$$

When the assumption in Method III is valid, i.e., when the partial derivatives in (5.1) cannot change sign within the limits of the uncertainties, then the Δx_k^{III} represent the true uncertainties, i.e.,

$$\Delta x_k^{III} = \Delta x_k, \qquad k = 1, 2, \ldots, n, \tag{5.12}$$

when the assumption in (5.1) is valid.

5.3 Intervals of Uncertainty in Unknowns

Let us now denote the intervals of uncertainty in the unknowns in (5.10) as given by Method III by U_k^{III}, $k = 1, 2, \ldots, n$, i.e.,

$$U_k^{III} = [x_k + e_{kk}, x_k + d_{kk}] \qquad k = 1, 2, \ldots, n, \tag{5.13}$$

(see (1.12)).
Then,

$$w(U_k^{III}) = d_{kk} - e_{kk}, \qquad k = 1, 2, \ldots, n. \tag{5.14}$$

Now, when the assumption in Method III is valid, the intervals in (5.10) represent the true intervals of uncertainty, i.e.,

$$e_k = -e_{kk}, \qquad d_k = d_{kk}, \qquad k = 1, 2, \ldots, n, \tag{5.15}$$

(see (1.12)).
Thus:

Subject to the assumption for Method III in (5.1),
$$U_k^{III} = U_k, \qquad k = 1, 2, \ldots, n. \tag{5.16}$$

And it clearly follows that:

Subject to the assumption for Method III in (5.1),
$$w(U_k^{III}) = w(U_k), \qquad k = 1, 2, \ldots, n. \tag{5.17}$$

At this stage, we conveniently consider for comparison the intervals of uncertainty U_k^I, say, given by Method I, and the intervals of uncertainty U_k^{II}, say, given by Method II.

Intervals of Uncertainty by Method I

We denote the intervals of uncertainty given by Method I by

$$U_k^I = [x_k - \Delta x_k^I, x_k + \Delta x_k^I], \qquad k = 1, 2, \ldots, n. \tag{5.18}$$

Hence, for each of $k = 1, 2, \ldots, n$ we have by (3.15) that
$$U_k^I \supseteq [x_k - \Delta x_k, x_k + \Delta x_k]$$
$$\supseteq U_k \qquad \text{by (1.15).}$$

Thus,

$$U_k \subseteq U_k^I, \qquad k = 1, 2, \ldots, n. \tag{5.19}$$

And further, we clearly have that

$$w(U_k^I) = 2\Delta x_k^I \geq 2\Delta x_k \geq w(U_k), \qquad k = 1, 2, \ldots, n, \tag{5.20}$$

(see (1.17)).

Intervals of Uncertainty by Method II

Similarly, we denote the intervals of uncertainty given by Method II by

$$U_k^{II} = [x_k - \Delta x_k^{II}, x_k + \Delta x_k^{II}], \qquad k = 1, 2, \ldots, n. \tag{5.21}$$

Hence, for each of $k = 1, 2, \ldots, n$ we have

$$w(U_k^{II}) = 2\Delta x_k^{II}$$
$$\approx d_{kk} - e_{kk}, \qquad \text{by (4.58), (4.59), and (4.60)}$$
$$= w(U_k^{III}), \qquad \text{by (5.14).}$$

Thus,

$$w(U_k^{II}) \approx w(U_k^{III}), \qquad k = 1, 2, \ldots, n. \tag{5.22}$$

Hence, we have by (5.17) that:

Subject to the assumption in (5.1),

$$w(U_k^{II}) \approx w(U_k), \qquad k = 1, 2, \ldots, n. \tag{5.23}$$

But let us note that although the widths of the U_k^{II} are nearly equal to those of the U_k, the corresponding intervals do not coincide that closely.

For compare the lower endpoints, say, of the intervals in (5.21) with those in (1.12). We then have that the difference between the lower endpoints in the kth intervals, say, is

$$(x_k - e_k) - (x_k - \Delta x_k^{II}) = \Delta x_k^{II} - e_k$$
$$= \Delta x_k^{II} + e_{kk}, \qquad \text{by (5.15),}$$
$$= -\sum_{j=1}^{n} c_{kj} e_{jk} \, \text{sign}(x_j), \qquad \text{by (4.59).}$$

Let us now return to our consideration of Method III.

5.4 Example

We now determine by Method III the uncertainties in the unknowns for the system of equations in (1.43) for the uncertainties in the coefficients and constants in (1.44).

To determine the increase in x_1 we require by (5.4), (1.46), and (1.47) to solve the system of equations

$$(0\cdot53 + \varepsilon)x_1^* + (0\cdot86 - \varepsilon)x_2^* + (0\cdot48 + \varepsilon)x_3^* = 0\cdot64 - \varepsilon$$
$$(0\cdot94 - \varepsilon)x_1^* + (-0\cdot47 + \varepsilon)x_2^* + (0\cdot85 - \varepsilon)x_3^* = 3\cdot12 + \varepsilon \quad (5.24)$$
$$(0\cdot87 - \varepsilon)x_1^* + (0\cdot55 + \varepsilon)x_2^* + (0\cdot26 - \varepsilon)x_3^* = 0\cdot83 + \varepsilon$$

where $\varepsilon = 0\cdot01$.

We obtain

$$x_1^* = 1\cdot1178, \qquad x_2^* = -1\cdot0668, \qquad x_3^* = 1\cdot9045. \quad (5.25)$$

Therefore, by (5.5) and (1.47) the maximum possible increase in x_1 is

$$x_1^* - x_1 = 1\cdot1178 - 0\cdot9730 = 0\cdot1448, \quad (5.26)$$

and we note that this value is equal to d_{11} given in (4.69) (see (5.7)).

Then, solving the system of equations in (5.24) with the sign of each ε changed we obtain

$$x_1^* = 0\cdot8325, \qquad x_2^* = -0\cdot9276, \qquad x_3^* = 2\cdot1790 \quad (5.27)$$

so that by (1.47) the maximum decrease in x_1 is

$$0\cdot8325 - 0\cdot9730 = -0\cdot1405. \quad (5.28)$$

And this is, of course, equal to e_{11} in (4.69).

Thus, by (1.47), (5.26), and (5.28) the estimated interval of uncertainty in x_1 is

$$[0\cdot9730 - 0\cdot1405, 0\cdot9730 + 0\cdot1448]. \quad (5.29)$$

And if the assumption in Method III is valid for the example then

$$x_1^* \in [0\cdot9730 - 0\cdot1405, 0\cdot9730 + 0\cdot1448]. \quad (5.30)$$

And by (5.11), (5.26), and (5.28) the estimate of the uncertainty by Method III in x_1 is

$$\Delta x_1^{III} = \max(|-0\cdot1405|, |0\cdot1448|) = 0\cdot1448. \quad (5.31)$$

The procedure has now to be repeated for Δx_2^{III} and Δx_3^{III}.

We should note that it is poor numerical practice to obtain a result as the difference between two nearly equal numbers if a method for finding the result directly is available. Our determination of the uncertainties in the unknowns as the difference between two solutions is satisfactory, however, provided each solution satisfies the given accuracy requirements. For the accuracy requirements for the uncertainties are not greater in the absolute sense than the accuracy requirements for the solution itself, i.e., the uncertainties are not required to more decimal places than the number of decimal places required in the solution.

In general, however, we would recommend finding the uncertainties in the unknowns directly by determining the columns of **D** and **E** (see (4.68)) rather than by determining the differences between solution vectors.

5.5 Validity of Assumption in Method III

We seek sufficient conditions to ensure that the assumption made in Method III is valid, namely, to ensure that the partial derivatives in (5.1) cannot change sign within the limits of the uncertainties in the coefficients and constants.

Now, by (4.11) and (4.12) the partial derivatives in (5.1) are given by

$$\frac{\partial x_k^*}{\partial a_{ij}^*} = -x_j^* b_{ki}^*, \qquad \frac{\partial x_k^*}{\partial c_i^*} = b_{ki}^*, \qquad i, j, k = 1, 2, \ldots, n, \quad (5.32)$$

where the a_{ij}^* and c_i^* are given by (1.7) and where, corresponding to a given set of values of a_{ij}^* and c_i^*, we have the unknowns x_k^* and the elements of the inverse b_{ki}^*.

Hence, the partial derivatives cannot change sign within the intervals of uncertainty in the coefficients and constants if the

$$x_j^* \quad \text{and} \quad b_{ki}^*, \qquad i, j, k = 1, 2, \ldots, n, \qquad (5.33)$$

cannot change sign.

But

$$x_j^* \in [x_j - \Delta x_j, x_j + \Delta x_j], \qquad b_{ki}^* \in [b_{ki} - \Delta b_{ki}, b_{ki} + \Delta b_{ki}], \\ i, j, k = 1, 2, \ldots, n \qquad (5.34)$$

(see (1.11), (1.12), and (1.15)).

Hence, the x_j^* and the b_{ki}^* cannot change sign if the intervals in (5.34) do not contain zero as an interior point, i.e., if

$$\Delta x_j \le |x_j|, \qquad \Delta b_{ki} \le |b_{ki}|, \qquad i, j, k = 1, 2, \ldots, n. \qquad (5.35)$$

Thus, (5.35) represents sufficient conditions to ensure that the partial derivatives in (5.32) cannot change sign within the limits of the given uncertainties.

In particular, we have by (5.1), (5.32), and (5.35) that:

> The assumption in Method III that the partial derivatives cannot change sign is valid in so far as the calculation of the interval of uncertainty for x_k is concerned if
>
> $$\Delta x_j \le |x_j|, \qquad \Delta b_{ki} \le |b_{ki}|, \qquad i, j = 1, 2, \ldots, n$$
>
> (5.36)

(see (5.3) and (5.8)).

Now, the estimates of the uncertainties by Method I are upper bounds of the uncertainties (see (3.15) and (3.31)). And it will be seen in (6.18) and (6.66) that the estimates by Method IV are also upper bounds. Thus

$$\Delta x_i^{\mathrm{I}} \geq \Delta x_i, \qquad \Delta b_{ij}^{\mathrm{I}} \geq \Delta b_{ij}, \qquad i, j = 1, 2, \ldots, n,$$

and

$$\Delta x_i^{\mathrm{IV}} \geq \Delta x_i, \qquad \Delta b_{ij}^{\mathrm{IV}} \geq \Delta b_{ij}, \qquad i, j = 1, 2, \ldots, n.$$

Hence,

if $\Delta x_i^{\mathrm{I}} \leq |x_i|$ or $\Delta x_i^{\mathrm{IV}} \leq |x_i|$ then $\Delta x_i \leq |x_i|$, $i = 1, 2, \ldots, n,$

and

if $\Delta b_{ki}^{\mathrm{I}} \leq |b_{ki}|$ or $\Delta b_{ki}^{\mathrm{IV}} \leq |b_{ki}|$ then $\Delta b_{ki} \leq |b_{ki}|,$

$$i, k = 1, 2, \ldots, n.$$

Hence, we have by (5.36) the following sufficient conditions for the assumption in Method III to be valid as far as calculating the interval of uncertainty in x_k is concerned:

$$\Delta x_i^{\mathrm{I}} \leq |x_i| \qquad \text{or} \qquad \Delta x_i^{\mathrm{IV}} \leq |x_i|, \qquad i = 1, 2, \ldots, n, \quad (5.37)$$

and

$$\Delta b_{ki}^{\mathrm{I}} \leq |b_{ki}| \qquad \text{or} \qquad \Delta b_{ki}^{\mathrm{IV}} \leq |b_{ki}|, \qquad i = 1, 2, \ldots, n. \quad (5.38)$$

Then Method III gives the true interval of uncertainty in x_k and $\Delta x_k^{\mathrm{III}} = \Delta x_k$.

5.6   A Computational Note on Method III

We have seen that the diagonal elements of the matrices $\mathbf{D}$ and $\mathbf{E}$ lead to the intervals of uncertainty and the uncertainties in the unknowns by Method III (see (5.7), (5.9), (5.10), and (5.11)).

We now therefore consider the solution procedures for the $2n$ systems of equations to obtain the $2n$ columns of $\mathbf{D}$ and $\mathbf{E}$, the systems of equations to obtain the kth columns of $\mathbf{D}$ and $\mathbf{E}$ being given in (4.53) and (4.55). The system of equations to obtain the kth column of $\mathbf{D}$ is thus

$$d_{mk} = \sum_{i=1}^{n} \sum_{j=1}^{n} \varepsilon_{ij} |x_j| b_{mi} \operatorname{sign}(b_{ki}) + \sum_{i=1}^{n} \varepsilon_i b_{mi} \operatorname{sign}(b_{ki})$$

$$+ \sum_{i=1}^{n} \sum_{j=1}^{n} \varepsilon_{ij} d_{jk} b_{mi} \operatorname{sign}(x_j b_{ki}), \qquad m = 1, 2, \ldots, n. \quad (5.39)$$

We may recall that (5.39) is merely the basic equation of differentials

$$\delta \mathbf{x} = \mathbf{B}\, \delta \mathbf{c} - \mathbf{B}(\delta \mathbf{A})\mathbf{x} - \mathbf{B}(\delta \mathbf{A})\, \delta \mathbf{x} \qquad (5.40)$$

(see (3.4)) with the notation

$$\delta \mathbf{x} = (d_{1k}, d_{2k}, \ldots, d_{nk})'$$

and with $\delta \mathbf{A} = (\delta a_{ij})$, $\delta \mathbf{c} = (\delta c_i)$ where

$$\delta a_{ij} = -\varepsilon_{ij}\, \mathrm{sign}(x_j b_{ki}), \qquad \delta c_i = \varepsilon_i\, \mathrm{sign}(b_{ki}), \qquad i, j = 1, 2, \ldots, n \qquad (5.41)$$

(see (5.3), and (4.50) to (4.53)).

And it is the value of d_{kk} that we require in (5.39) (see (5.7)).

Now, (5.39) is a system of n linear algebraic equations for the n unknowns $d_{1k}, d_{2k}, \ldots, d_{nk}$.

Hence, we can use Gaussian elimination to solve the system of equations.

Further, a little consideration will show that (5.39) is a strongly diagonal system of equations because the coefficients of the d_{jk} $(j = 1, 2, \ldots, n)$ in the third term on the right-hand side of (5.39) are small, the ε's being of the first order of small quantities.

Hence, we can solve (5.39) by Jacobi or Gauss–Seidel iteration.

But let us consider the following iterative scheme based directly on the form in (5.39) to obtain the kth column of $\mathbf{D}$:

$$\begin{aligned}
d_{mk}^{(r)} = &\sum_{i=1}^{n} \sum_{j=1}^{n} \varepsilon_{ij} |x_j| b_{mi}\, \mathrm{sign}(b_{ki}) + \sum_{i=1}^{n} \varepsilon_i b_{mi}\, \mathrm{sign}(b_{ki}) \\
&+ \sum_{i=1}^{n} \sum_{j=1}^{n} \varepsilon_{ij} d_{jk}^{(p)} b_{mi}\, \mathrm{sign}(x_j b_{ki}), \qquad m = 1, 2, \ldots, n, \\
&\hspace{7cm} r = 1, 2, \ldots,
\end{aligned} \qquad (5.42)$$

in which we use the notation $d_{mk}^{(r)}$ to indicate the rth iterate of d_{mk}, and in which $d_{jk}^{(p)}$ in the third term on the right-hand side has still to be specified.

First, let us for convenience write (5.42) as

$$d_{mk}^{(r)} = f_{mk} + \sum_{i=1}^{n} \sum_{j=1}^{n} \varepsilon_{ij} d_{jk}^{(p)} b_{mi}\, \mathrm{sign}(x_j b_{ki}), \qquad \begin{aligned} m &= 1, 2, \ldots, n \\ r &= 1, 2, \ldots, \end{aligned} \qquad (5.43)$$

where

$$f_{mk} = \sum_{i=1}^{n} \sum_{j=1}^{n} \varepsilon_{ij} |x_j| b_{mi}\, \mathrm{sign}(b_{ki}) + \sum_{i=1}^{n} \varepsilon_i b_{mi}\, \mathrm{sign}(b_{ki}), \qquad (5.44)$$

i.e.,

$$f_{mk} = \sum_{i=1}^{n} (\varepsilon_i + \sum_{j=1}^{n} \varepsilon_{ij}|x_j|)b_{mi} \text{ sign}(b_{ki}). \tag{5.45}$$

And now let us specify $d_{jk}^{(p)}$ in the second term on the right-hand side of (5.43) by

$$d_{jk}^{(p)} = 0, \qquad r = 1, \quad j = m, m+1, \ldots, n, \tag{5.46}$$

$$d_{jk}^{(p)} = d_{jk}^{(r-1)}, \qquad r = 2, 3, \ldots, \quad j = m, m+1, \ldots, n, \tag{5.47}$$

$$d_{jk}^{(p)} = d_{jk}^{(r)}, \qquad r = 1, 2, \ldots, \quad j = 1, 2, \ldots, m-1; m \neq 1. \tag{5.48}$$

We thus use each new iterate as soon as it is obtained so that we have a Gauss–Seidel type procedure above.

Then it is expected that only a few iteration cycles would be required in practice to obtain convergence because the second term on the right-hand side of (5.43) containing the d's is of the second order of small quantities, both the ε's and the d's in this term being of the first order of small quantities.

Carrying out the procedure for $k = 1, 2, \ldots, n$ we can construct $\mathbf{D}$, the diagonal elements of $\mathbf{D}$ being required by (5.7). And an iterative scheme similar to that in (5.43) can be used to determine the elements of $\mathbf{E}$.

Thus, the volume of computation indicated above involves the iterative solution of $2n$ systems of equations, each solution consisting only of a few iterative cycles.

5.7 An Approximate Computational Procedure

The volume of computation indicated in Section 5.6 can be approximately halved if we estimate the diagonal elements of the matrix $\mathbf{E}$ from the approximate relation in (4.67):

$$\Delta x_k^{II} \approx \tfrac{1}{2}(d_{kk} - e_{kk}), \qquad k = 1, 2, \ldots, n. \tag{5.49}$$

For let us note by (5.45) and (4.21) that

$$f_{kk} = \sum_{i=1}^{n} (\varepsilon_i + \sum_{j=1}^{n} \varepsilon_{ij}|x_j|)|b_{ki}| = \Delta x_k^{II}. \tag{5.50}$$

Hence, if the iterative scheme in (5.43) can be regarded as having converged at the rth iterative cycle so that

$$d_{kk} = d_{kk}^{(r)} \tag{5.51}$$

then by (5.50) and (5.49)

$$e_{kk} \approx d_{kk}^{(r)} - 2f_{kk}. \tag{5.52}$$

Thus, the diagonal elements of $\mathbf{E}$ can be determined at the time that the diagonal elements of $\mathbf{D}$ are determined.

And in practice the estimates of the e_{kk} obtained in this way are sufficiently close to the true values of the e_{kk}, especially if the Δx_k^{II} agree closely with the diagonal elements of $\mathbf{D}$.

Let us now illustrate the above by means of our numerical example in (1.43) for an uncertainty of 0·01 in each coefficient and constant.

For the matrices $\mathbf{D}$ and $\mathbf{E}$ which were obtained independently of one another in (4.69) we have by (5.13) and (1.47) that

$$\begin{aligned}
U_1^{\mathrm{III}} &= [0{\cdot}9730 - 0{\cdot}1405, \qquad 0{\cdot}9730 + 0{\cdot}1448] \\
U_2^{\mathrm{III}} &= [-0{\cdot}9962 - 0{\cdot}0710, \quad -0{\cdot}9962 + 0{\cdot}0684]. \qquad (5.53) \\
U_3^{\mathrm{III}} &= [2{\cdot}0438 - 0{\cdot}2165, \qquad 2{\cdot}0438 + 0{\cdot}2222]
\end{aligned}$$

But suppose we use the diagonal elements of $\mathbf{D}$ in (4.69) and the Δx_k^{II} in (4.35) to estimate the diagonal elements of $\mathbf{E}$ by (5.49). Then, the approximate values of the lower endpoints of the U_k^{III} do not in fact differ by more than one unit in the fourth decimal place from the true values in (5.53).

Thus, the method of this section would lead to approximate results for our example which do not differ by more than one unit in the fourth decimal place from those in (5.53), the volume of computation to obtain the approximate results being half that to obtain the true results.

In the next section we consider the uncertainties in the elements of the inverse by Method III; the reader may at this stage go directly to Section 5.9 without loss of continuity.

5.8   Uncertainties in Elements of Inverse by Method III

To obtain the intervals of uncertainty in the elements of the inverse, we require to find for each element of the inverse the maximum increase and decrease for changes in the coefficients within the limits of their uncertainties. (Since the columns of the inverse can be regarded as the solution vectors when the right-hand sides are the unit vectors, only the uncertainties in the coefficients need be considered.)

Because the elements of the inverse correspond to the unknowns, the assumption in Method III in place of (5.1) is now that the partial derivatives of the elements of the inverse with respect to the coefficients do not change sign within the limits of the uncertainties in the coefficients. Thus our assumption is now that

$$\frac{\partial b^*_{mh}}{\partial a^*_{ij}}, \qquad m, h, i, j = 1, 2, \ldots, n, \tag{5.54}$$

do not change sign for the a^*_{ij} given in (1.7) where $\mathbf{B}^* = (b^*_{mh})$ is the inverse of $\mathbf{A}^* = \mathbf{A} + \delta\mathbf{A}$ (the inverse of $\mathbf{A}$ being denoted by $\mathbf{B} = (b_{mh})$).

Let us now determine in particular the maximum increase in the element b_{kh}, say, for changes in the coefficients within the limits of the uncertainties, subject to the assumption in Method III being valid.

In the basic equation of differentials (5.40)

$$\delta\mathbf{x} = \mathbf{B}\,\delta\mathbf{c} - \mathbf{B}(\delta\mathbf{A})\mathbf{x} - \mathbf{B}(\delta\mathbf{A})\,\delta\mathbf{x} \tag{5.55}$$

we must put

$$\mathbf{x} = (b_{1h}, b_{2h}, \ldots, b_{nh})'$$

(i.e., the hth column of $\mathbf{B}$) and hence

$$\delta\mathbf{x} = (\delta b_{1h}, \delta b_{2h}, \ldots, \delta b_{nh})'.$$

And we must put

$$\delta\mathbf{c} = \mathbf{0}$$

because here $\mathbf{c}$ refers to the hth unit vector used to obtain the hth column of the inverse, i.e., $\mathbf{c}$ refers to $(0, 0, \ldots, 0, 1, 0, \ldots, 0)'$ with the unit element in the hth position.

Then (5.55) becomes

$$\delta b_{mh} = -\sum_{i=1}^{n}\sum_{j=1}^{n} b_{mi}\,\delta a_{ij} b_{jh} - \sum_{i=1}^{n}\sum_{j=1}^{n} b_{mi}\,\delta a_{ij}\,\delta b_{jh}, \qquad m = 1, 2, \ldots, n \tag{5.56}$$

(compare with (4.50)).

And to obtain the maximum increase in b_{kh} it is clear that the set of changes in (5.3)

$$\delta a_{ij} = \varepsilon_{ij}\,\mathrm{sign}(-x_j b_{ki}) = -\varepsilon_{ij}\,\mathrm{sign}(x_j b_{ki}) \qquad i, j = 1, 2, \ldots, n, \tag{5.57}$$

becomes

$$\delta a_{ij} = -\varepsilon_{ij}\,\mathrm{sign}(b_{jh} b_{ki}), \qquad i, j = 1, 2, \ldots, n, \tag{5.58}$$

x_j in (5.57) being replaced by b_{jh}.

Then, on solving (5.56) with the δa_{ij} in (5.58), the maximum increase in the element b_{kh} on the basis of the assumption in Method III is given by δb_{kh}.

Hence, to obtain the $2n^2$ endpoints of the intervals of uncertainty of the n^2 elements of the inverse requires the solution of $2n^2$ systems of equations of the form in (5.56). This clearly involves a considerable volume of computation, although the iterative and approximate procedures given in Sections 5.6 and 5.7 can be used.

For example, suppose we wish to determine the maximum increase in the element b_{kh} for changes in the coefficients within the limits of their uncertainties.

Then, corresponding to (5.42) we have by (5.56) and (5.58)

$$\delta b_{mh}^{(r)} = \sum_{i=1}^{n} \sum_{j=1}^{n} \varepsilon_{ij} |b_{jh}| |b_{mi}| \operatorname{sign}(b_{ki}) + \sum_{i=1}^{n} \sum_{j=1}^{n} \varepsilon_{ij} b_{mi} \, \delta b_{jh}^{(p)} \operatorname{sign}(b_{jh} b_{ki}), \quad (5.59)$$

$$m = 1, 2, \ldots, n.$$

And we can implement the Gauss–Seidel type iterative procedure with specifications similar to those in (5.46), (5.47), and (5.48).

Then, when convergence is achieved, $\delta b_{kh}^{(r)}$ gives the maximum increase in the element b_{kh}. And the maximum decrease in b_{kh} can be similarly obtained, an approximation to this maximum decrease being found with a trivial amount of computation by adopting the procedure of Section 5.7.

Next, let us determine sufficient conditions to ensure that the interval of uncertainty by Method III in the element b_{kh} of the inverse is the true interval of uncertainty.

Corresponding to (5.37) and (5.38), we have

$$\Delta b_{ih}^{\mathrm{I}} \le |b_{ih}| \quad \text{or} \quad \Delta b_{ih}^{\mathrm{IV}} \le |b_{ih}|, \qquad i = 1, 2, \ldots, n, \qquad (5.60)$$

and

$$\Delta b_{ki}^{\mathrm{I}} \le |b_{ki}| \quad \text{or} \quad \Delta b_{ki}^{\mathrm{IV}} \le |b_{ki}| \qquad i = 1, 2, \ldots, n. \qquad (5.61)$$

Thus, the conditions in (5.60) and (5.61) ensure that the interval of uncertainty in b_{kh} by Method III is the true interval of uncertainty. And in general we have:

Method III gives the true intervals of uncertainty and the true uncertainties in all the elements of the inverse

$$\text{if} \quad \Delta b_{ij}^{\mathrm{I}} \le |b_{ij}|, \qquad i, j = 1, 2, \ldots, n, \qquad (5.62)$$
$$\text{or} \quad \Delta b_{ij}^{\mathrm{IV}} \le |b_{ij}|, \qquad i, j = 1, 2, \ldots, n.$$

Either set of conditions in (5.62) will ensure that

$$\frac{\partial b_{kh}^{*}}{\partial a_{ij}^{*}} = -b_{jh}^{*} b_{ki}^{*}, \qquad k, h, i, j = 1, 2, \ldots, n, \qquad (5.63)$$

(see (5.54) and (5.32)) cannot change sign within the limits of the uncertainties in the coefficients. This then ensures that the assumption in Method III is valid for computing the intervals of uncertainty in the elements of the inverse.

And it clearly follows from (5.62) that when the elements of the inverse are large in magnitude then Method III may be expected to give the true intervals of uncertainty and the true uncertainties in the elements of the inverse.

5.9 Example on Validity of Assumption in Method III

Let us examine whether the conditions of application of Method III in (5.37) and (5.38) are satisfied for our example in (1.43) and (1.44). By (3.25)

$$\Delta x_i^{\mathrm{I}} = 0.3593, \qquad i = 1, 2, 3,$$

while by (1.47)

$$x_1 = 0.9730, \qquad x_2 = -0.9962, \qquad x_3 = 2.0438. \qquad (5.64)$$

Hence,

$$\Delta x_i^{\mathrm{I}} = 0.3593 < |x_i|, \qquad i = 1, 2, 3, \qquad (5.65)$$

so that the unknowns cannot change sign within their intervals of uncertainty and (5.37) is satisfied.

But by (3.35) we have

$$\Delta \mathbf{B}^{\mathrm{I}} = \begin{bmatrix} 0.2509 & 0.1238 & 0.2866 \\ 0.2509 & 0.1238 & 0.2866 \\ 0.2509 & 0.1238 & 0.2866 \end{bmatrix}, \qquad (5.66)$$

while by (1.46)

$$\mathbf{B} = \begin{bmatrix} -1.0574 & 0.0724 & 1.7153 \\ 0.8878 & -0.5017 & 0.0013 \\ 1.6603 & 0.8189 & -1.8963 \end{bmatrix}. \qquad (5.67)$$

Hence, the first set of conditions in (5.38) is not satisfied because

$$\Delta b_{12}^{\mathrm{I}} = 0.1238 \nleqslant 0.0724 = |b_{12}|$$

and

$$\Delta b_{23}^{\mathrm{I}} = 0.2866 \nleqslant 0.0013 = |b_{23}|. \qquad (5.68)$$

In applying the alternative set of conditions in (5.38) we have by (6.73) that

$$\Delta \mathbf{B}^{\mathrm{IV}} = \begin{bmatrix} 0.1122 & 0.0434 & 0.1125 \\ 0.0549 & 0.0212 & 0.0550 \\ 0.1726 & 0.0667 & 0.1730 \end{bmatrix}. \qquad (5.69)$$

While the uncertainties in (5.69) are all less than those in (5.66) we nevertheless find that the second set of conditions in (5.38) is still not satisfied because

$$\Delta b_{23}^{\mathrm{IV}} = 0.0550 \nleqslant 0.0013 = |b_{23}|. \qquad (5.70)$$

And indeed a lower bound of the uncertainty Δb_{23} is

$$\Delta b_{23}^{V} = (1 - 2 \times 0.08611) \times 0.0550 = 0.0455$$

(see (7.18), (7.29), (5.69), and (1.48)) so that we have

$$\Delta b_{23} \geq \Delta b_{23}^{V} = 0.0455 > 0.0013 = |b_{23}|, \qquad \text{i.e.,} \qquad \Delta b_{23} > |b_{23}|. \tag{5.71}$$

Hence, we may well expect that b_{23} can change sign within the limits of the uncertainties in the coefficients, it being recalled that the conditions in (5.38) are only sufficient ones.

Hence, by (5.32) it would follow that for $k = 2$ the interval in (5.10) may not be the true interval of uncertainty in x_2.

And indeed the interval of uncertainty in x_2 by Method III

$$U_2^{\text{III}} = [x_2 + e_{22}, x_2 + d_{22}] = [-0.9962 - 0.0710, -0.9962 + 0.0684]$$
$$= [-1.0672, -0.9278] \tag{5.72}$$

is not the true interval of uncertainty in x_2 for our example (see (5.13), (5.64), and (4.69)).

For when we examine the rows of $\mathbf{D}$ and $\mathbf{E}$ in (4.69)

$$\mathbf{D} = \begin{bmatrix} 0.1448 & 0.0288 & -0.1371 \\ -\overline{0.0706} & 0.0684 & 0.0195 \\ -0.1393 & -\overline{0.0519} & 0.2222 \end{bmatrix},$$

$$\mathbf{E} = \begin{bmatrix} -0.1405 & -0.0299 & 0.1366 \\ \overline{0.0686} & -0.0710 & -0.0190 \\ 0.1352 & \overline{0.0539} & -0.2165 \end{bmatrix}, \tag{5.73}$$

it is clear that not all of the elements in the second rows of $\mathbf{D}$ and $\mathbf{E}$ lie in the interval

$$[e_{22}, d_{22}]. \tag{5.74}$$

In particular

$$e_{21} \notin [e_{22}, d_{22}] \quad \text{because} \quad e_{21} = 0.0686 > 0.0684 = d_{22}. \tag{5.75}$$

But the elements of the second rows of $\mathbf{D}$ and $\mathbf{E}$ all represent changes in x_2 for changes in the coefficients and constants within the limits of the uncertainties. Hence, one possible value of x_2^*, i.e., one possible true value of x_2, is

$$x_2^* = x_2 + e_{21} = -0.9962 + 0.0686 = -0.9276. \tag{5.76}$$

Hence, the interval in (5.72) is not the true interval of uncertainty in x_2 because we have shown that at least one value of x_2^* lies outside it.

Thus, the assumption in (5.1) must fail as far as the interval of uncertainty in x_2 is concerned.

Clearly, there is a likelihood that the conditions in (5.37) and (5.38) may not be satisfied if the absolute values of the unknowns and of the elements of the inverse are small (see (5.70) or (5.71), for example). And under these circumstances, then, there is a possibility that the assumption in (5.1) may not hold.

Let us now note in passing that:

> In those cases where Method III does not lead to the true intervals of uncertainty it leads to intervals con-   (5.77)
> tained in the true intervals.

For the intervals given by Method III always correspond to changes in the unknowns that can occur within the limits of the uncertainties in the coefficients and right-hand constants, so the intervals given by Method III cannot fall outside the true intervals of uncertainty.

Thus, we have that in general

$$U_k^{III} = [x_k + e_{kk}, x_k + d_{kk}] \subseteq U_k, \qquad k = 1, 2, \ldots, n \qquad (5.78)$$

(see (5.13) and also (5.12)). But let us note that there is the tacit assumption in (5.78) that the e_{kk} are negative and the d_{kk} are positive, even when the assumption in (5.1) is not valid (see (5.7), (5.9), and (5.10)).

5.10 Conclusion

Sufficient conditions of application for Method III are given in (5.37) and (5.38). These conditions can be expected to be satisfied when the unknowns and the elements of the inverse are not small in magnitude, assuming that the given uncertainties in the coefficients and constants are not too large.

Method III gives the true intervals of uncertainty and the true uncertainties in the unknowns and in the elements of the inverse when the conditions of application are satisfied.

And when Method III is applicable the volume of computation is of the same order of magnitude as in the alternative linear programming approach (Method IX). Method III is the more direct, however, and perhaps for this reason it is to be preferred to the linear programming approach.

But where the assumption in (5.1) cannot be shown to be valid, the linear programming approach will always lead to the true intervals of uncertainty when these exist. However, the volume of computation may now become very large.

In the next chapter (Method IV) we show how to obtain upper bounds of the uncertainties and hence intervals containing the true intervals of uncertainty. It will be clear that Method IV involves less computation than Method III and that Method IV may be a suitable approach when the assumption in (5.1) is not valid, i.e., when Method III is not applicable. For it will be seen that the intervals given by Method IV are often not much wider than the true intervals of uncertainty.

CHAPTER 6

Method IV: Uncertainties in Unknowns by Modulus Analysis

6.1 Introduction

In Method I we analysed the basic equation of differentials (3.4) using the property that the norm of the sum or difference of two matrices or two vectors is less than or equal to the sum of the norms, and that the norm of the product of two matrices is less than or equal to the product of the norms. We then obtained an upper bound for the uncertainty of largest magnitude.

In our present analysis we use the property that the modulus of the sum or difference of two numbers is less than or equal to the sum of the moduli, the modulus of the product of two numbers being equal to the product of the moduli. The present analysis leads to upper bounds for the individual uncertainties.

It is found in practice that the upper bounds given by Method IV are frequently very close to the true uncertainties in the unknowns, a situation which would generally hold when the elements of the inverse of the coefficient matrix are not large and the ε_{ij} and ε_i are small.

Further, it is usually found that the largest value given by Method IV is far less than the upper bound for the uncertainties given by Method I. That Method IV should give a finer and truer picture of the uncertainties than Method I might well be expected, since, loosely speaking, Method IV uses the properties of the elements of the matrices and vectors directly while Method I uses the properties of the matrices and vectors as a whole.

This might be a suitable point to clarify the two uses of the term 'uncertainty in an unknown'. The definition of the term is given in (1.14). But we also use the term 'uncertainty in an unknown' for the estimate of the true uncertainty by a particular method, the terms *uncertainty* and *true uncertainty* being used synonymously. Thus, when we are comparing the uncertainties given by two methods we are in fact comparing the estimates of the true uncertainty by the two methods.

6.2 Definition of Uncertainties by Method IV

In Method I we applied norm analysis to the basic equation of differentials (3.4). And in Method II we in effect neglected the term $B(\delta A)\,\delta x$ in the basic equation of differentials and applied modulus analysis to the first-order terms (see (4.15) and (4.17)).

In Method IV we now proceed as in Method II, but we include the second-order term as well in our analysis.

Writing the basic equation of differentials out in full,

$$\delta x_k = -\sum_{i=1}^{n}\sum_{j=1}^{n} b_{ki}\,\delta a_{ij}(x_j + \delta x_j) + \sum_{i=1}^{n} b_{ki}\,\delta c_i, \quad k = 1, 2, \ldots, n, \quad (6.1)$$

and taking moduli, we have

$$|\delta x_k| \le \sum_{i=1}^{n}\sum_{j=1}^{n} |b_{ki}|\,|\delta a_{ij}|\,(|x_j| + |\delta x_j|) + \sum_{i=1}^{n} |b_{ki}|\,|\delta c_i|, \quad k = 1, 2, \ldots, n. \quad (6.2)$$

We now restrict ourselves to the case where the changes in the co-efficients and constants are within the limits of their uncertainties, i.e.,

$$|\delta a_{ij}| \le \varepsilon_{ij}, \qquad |\delta c_i| \le \varepsilon_i, \qquad i, j = 1, 2, \ldots, n. \quad (6.3)$$

And let us assume that the system of equations is not critically ill-conditioned, so that the Δx_j $(j = 1, 2, \ldots, n)$ exist. Then, subject to (6.3) we have for the changes in the unknowns as determined by (6.1) that

$$|\delta x_j| \le \Delta x_j, \qquad j = 1, 2, \ldots, n. \quad (6.4)$$

Now, replacing the δa_{ij} by ε_{ij}, the δc_i by ε_i, and the δx_j by Δx_j in the right-hand sides of (6.2), it follows by (6.3) and (6.4) that

$$\sum_{i=1}^{n}\sum_{j=1}^{n} |b_{ki}|\varepsilon_{ij}(|x_j| + \Delta x_j) + \sum_{i=1}^{n} |b_{ki}|\varepsilon_i, \quad k = 1, 2, \ldots, n \quad (6.5)$$

are upper bounds of the right-hand sides of (6.2), the $|b_{ki}|$ and the $|x_j|$ being fixed quantities for a given problem.

But since (6.5) gives upper bounds of the right-hand sides of (6.2) subject to (6.3), it follows that the expressions in (6.5) must be greater than or equal to the least upper bounds of the left-hand sides of (6.2), again subject to (6.3). Now, for changes in the coefficients and constants within the limits of their uncertainties we have that the Δx_k are the least upper bounds of the $|\delta x_k|$. Hence,

$$\Delta x_k \le \sum_{i=1}^{n} \sum_{j=1}^{n} |b_{ki}| \varepsilon_{ij} (|x_j| + \Delta x_j) + \sum_{i=1}^{n} |b_{ki}| \varepsilon_i, \qquad (6.6)$$
$$k = 1, 2, \ldots, n,$$

i.e.,

$$\Delta x_k \le \sum_{i=1}^{n} (c_{ki} (|x_i| + \Delta x_i) |+ b_{ki}| \varepsilon_i), \qquad k = 1, 2, \ldots, n, \quad (6.7)$$

(see (1.38)).

Thus, by (4.22) we have

$$\Delta x_k \le \Delta x_k^{II} + \sum_{i=1}^{n} c_{ki} \Delta x_i, \qquad k = 1, 2, \ldots, n. \qquad (6.8)$$

Or in matrix notation we have

$$\Delta \mathbf{x} \le \Delta \mathbf{x}^{II} + \mathbf{C} \Delta \mathbf{x}, \qquad (6.9)$$

the inequality sign, of course, being interpreted elementwise so that (6.9) in fact represents the n inequalities in (6.8).

Let us now denote the estimates of the uncertainties by Method IV by Δx_k^{IV} (the superscript IV to indicate estimates by Method IV). And let us define the Δx_k^{IV} to be equal to the right-hand sides of (6.6) with the Δx_j in (6.6) replaced by Δx_j^{IV}. Thus, by definition, we have

$$\Delta x_k^{IV} = \sum_{i=1}^{n} \sum_{j=1}^{n} |b_{ki}| \varepsilon_{ij} (|x_j| + \Delta x_j^{IV}) + \sum_{i=1}^{n} |b_{ki}| \varepsilon_i, \qquad (6.10)$$
$$k = 1, 2, \ldots, n,$$

i.e.,

$$\Delta x_k^{IV} = \sum_{i=1}^{n} (c_{ki} (|x_i| + \Delta x_i^{IV}) + |b_{ki}| \varepsilon_i), \qquad k = 1, 2, \ldots, n, \quad (6.11)$$

i.e.,

$$\Delta x_k^{IV} = \Delta x_k^{II} + \sum_{i=1}^{n} c_{ki} \Delta x_i^{IV}, \qquad k = 1, 2, \ldots, n, \qquad (6.12)$$

by (4.22).

And we may write (6.12) in matrix notation as

$$\Delta \mathbf{x}^{IV} = \Delta \mathbf{x}^{II} + \mathbf{C} \Delta \mathbf{x}^{IV} \qquad (6.13)$$

where

$$\Delta \mathbf{x}^{IV} = (\Delta x_k^{IV}). \qquad (6.14)$$

Hence,

$$(\mathbf{I} - \mathbf{C}) \Delta \mathbf{x}^{IV} = \Delta \mathbf{x}^{II} \qquad (6.15)$$

so that

$$\Delta \mathbf{x}^{IV} = (\mathbf{I} - \mathbf{C})^{-1} \Delta \mathbf{x}^{II} \qquad (6.16)$$

provided $(\mathbf{I} - \mathbf{C})$ is nonsingular, i.e., provided

$$\det(\mathbf{I} - \mathbf{C}) \neq 0. \tag{6.17}$$

Now, since the right-hand sides of (6.6) are upper bounds of the uncertainties Δx_k ($k = 1, 2, \ldots, n$), we might well expect from the definition in (6.10) that the Δx_k^{IV} are upper bounds as well.

Indeed, we prove in the next section that:

$$\text{If} \quad \rho(\mathbf{C}) < 1 \quad \text{then} \quad \Delta\mathbf{x}^{\mathrm{IV}} \geq \Delta\mathbf{x}. \tag{6.18}$$

And by (2.24) and (2.37) we have that $\rho(\mathbf{C}) < 1$ if

$$\phi_{\mathrm{norm}} \equiv \|\mathbf{C}\|_{\mathrm{I}} < 1 \quad \text{or if} \quad \|\mathbf{C}\|_{\mathrm{II}} < 1.$$

Hence the result in (6.18) includes the following result:

$$\text{If} \quad \phi_{\mathrm{norm}} \equiv \|\mathbf{C}\|_{\mathrm{I}} < 1 \quad \text{or} \quad \|\mathbf{C}\|_{\mathrm{II}} < 1 \quad \text{then} \quad \Delta\mathbf{x}^{\mathrm{IV}} \geq \Delta\mathbf{x}. \tag{6.19}$$

6.3   Conditions for the Δx_k^{IV} to be Upper Bounds

We now set out to prove the result in (6.18).

Let us assume that

$$\rho(\mathbf{C}) < 1. \tag{6.20}$$

Then, by (2.36) the system of equations is not critically ill-conditioned and the Δx_j exist (see (6.4)). Hence, if (6.20) holds then the relations (6.6) to (6.9) hold, these relations being all different forms of the same result. In particular (6.9) holds:

$$\Delta\mathbf{x} \leq \Delta\mathbf{x}^{\mathrm{II}} + \mathbf{C}\,\Delta\mathbf{x}. \tag{6.21}$$

Let us now denote the right-hand side of (6.21) by

$$\Delta\mathbf{x}^{\alpha_1}$$

and let us write

$$\Delta\mathbf{x}^{\alpha_0} \equiv \Delta\mathbf{x}. \tag{6.22}$$

Then, with the new notation we have from (6.21) that

$$\Delta\mathbf{x}^{\alpha_1} \geq \Delta\mathbf{x}^{\alpha_0} = \Delta\mathbf{x} \tag{6.23}$$

where

$$\Delta\mathbf{x}^{\alpha_1} = \Delta\mathbf{x}^{\mathrm{II}} + \mathbf{C}\,\Delta\mathbf{x}^{\alpha_0}. \tag{6.24}$$

And now let us replace $\Delta\mathbf{x}^{\alpha_0}$ on the right-hand side of (6.24) by $\Delta\mathbf{x}^{\alpha_1}$ and denote the new right-hand side by $\Delta\mathbf{x}^{\alpha_2}$. Thus,

$$\Delta\mathbf{x}^{\alpha_2} = \Delta\mathbf{x}^{\mathrm{II}} + \mathbf{C}\,\Delta\mathbf{x}^{\alpha_1}. \tag{6.25}$$

But since both $\mathbf{C}$ and $\Delta\mathbf{x}$ are nonnegative, it follows by (6.23) that $\mathbf{C}\,\Delta\mathbf{x}^{\alpha_1} \geq \mathbf{C}\,\Delta\mathbf{x}^{\alpha_0}$. Hence we have by (6.25) and (6.24) that

$$\Delta\mathbf{x}^{\alpha_2} \geq \Delta\mathbf{x}^{\alpha_1}. \tag{6.26}$$

And combining (6.23) and (6.26) we have

$$\Delta\mathbf{x}^{\alpha_2} \geq \Delta\mathbf{x}^{\alpha_1} \geq \Delta\mathbf{x}^{\alpha_0} = \Delta\mathbf{x}. \tag{6.27}$$

On repeating this procedure $(p-2)$ more times we have

$$\Delta\mathbf{x}^{\alpha_p} \geq \Delta\mathbf{x}^{\alpha_{p-1}} \geq \ldots \geq \Delta\mathbf{x}^{\alpha_1} \geq \Delta\mathbf{x}^{\alpha_0} = \Delta\mathbf{x} \tag{6.28}$$

where

$$\Delta\mathbf{x}^{\alpha_p} = \Delta\mathbf{x}^{\mathrm{II}} + \mathbf{C}\,\Delta\mathbf{x}^{\alpha_{p-1}}. \tag{6.29}$$

To prove the result in (6.18), we now prove that

$$\Delta\mathbf{x}^{\mathrm{IV}} \text{ exists if } \rho(\mathbf{C}) < 1 \tag{6.30}$$

and then that

$$\Delta\mathbf{x}^{\mathrm{IV}} - \Delta\mathbf{x}^{\alpha_p} \to \mathbf{0} \quad \text{as} \quad p \to \infty \quad \text{if} \quad \rho(\mathbf{C}) < 1. \tag{6.31}$$

Then (6.18) clearly follows by (6.28).

Now it is easy to see that

$$\det(\mathbf{I} - \mathbf{C}) \neq 0 \quad \text{if} \quad \rho(\mathbf{C}) < 1. \tag{6.32}$$

For the eigenvalues of any square matrix $\mathbf{S}$ are those values of λ for which $\det(\mathbf{S} - \lambda\mathbf{I}) = 0$. Hence

$$\det(\mathbf{S} - \mathbf{I}) \neq 0 \quad \text{if} \quad \lambda \neq 1 \quad \text{and hence if} \quad \rho(\mathbf{S}) = \max_i(|\lambda_i|) < 1. \tag{6.33}$$

But

$$\det(\mathbf{I} - \mathbf{S}) \neq 0 \quad \text{if} \quad \det(\mathbf{S} - \mathbf{I}) \neq 0. \tag{6.34}$$

For

$$\det(\mathbf{I} - \mathbf{S}) = \det(\mathbf{S} - \mathbf{I})$$

if $\mathbf{I} - \mathbf{S}$ is of even order and

$$\det(\mathbf{I} - \mathbf{S}) = -\det(\mathbf{S} - \mathbf{I})$$

if $\mathbf{I} - \mathbf{S}$ is of odd order. Hence for any square matrix $\mathbf{S}$

$$\det(\mathbf{I} - \mathbf{S}) \neq 0 \quad \text{if} \quad \rho(\mathbf{S}) < 1 \tag{6.35}$$

(see (2.32)). And in particular (6.32) holds.

Further, the condition $\rho(\mathbf{C}) < 1$ in (6.20) ensures by (2.36) that $\Delta\mathbf{x}^{\mathrm{II}}$ in (6.16) exists.

It now follows by (6.32) and (6.17) that (6.30) holds:

$$\Delta \mathbf{x}^{IV} \text{ exists being given by (6.16)} \qquad \text{if} \quad \rho(\mathbf{C}) < 1. \qquad (6.36)$$

Next to prove (6.31) we subtract (6.29) from (6.13) obtaining

$$\Delta \mathbf{x}^{IV} - \Delta \mathbf{x}^{\alpha_p} = \mathbf{C}(\Delta \mathbf{x}^{IV} - \Delta \mathbf{x}^{\alpha_{p-1}}).$$

On repeated application of this recurrence relation, we obtain

$$\Delta \mathbf{x}^{IV} - \Delta \mathbf{x}^{\alpha_p} = \mathbf{C}^p(\Delta \mathbf{x}^{IV} - \Delta \mathbf{x}^{\alpha_0}). \qquad (6.37)$$

But

$$\mathbf{C}^p \to \mathbf{0} \qquad \text{as} \quad p \to \infty \qquad (6.38)$$

by (2.17), (2.23), and (6.20). Hence by (6.37)

$$\Delta \mathbf{x}^{IV} - \Delta \mathbf{x}^{\alpha_p} \to \mathbf{0} \qquad \text{as} \quad p \to \infty,$$

i.e.,

$$\Delta \mathbf{x}^{\alpha_p} \to \Delta \mathbf{x}^{IV} \qquad \text{as} \quad p \to \infty. \qquad (6.39)$$

Hence (6.18) follows by (6.28).

And in the case when the uncertainties in the coefficients are all equal to ε we have that $\rho(\mathbf{C}) = \phi$ (see (2.46)). Hence, in this case (6.18) takes the form:

When all the $\varepsilon_{ij} = \varepsilon$ then $\Delta \mathbf{x}^{IV} \geq \Delta \mathbf{x}$ if $\phi < 1.$ (6.40)

6.4 Special Cases

In the special cases being considered, we determine simple expressions for the uncertainties by Method IV.

1. Uncertainties in Coefficients and Constants all Equal

We now consider the case when the uncertainties in the coefficients and constants are all equal to ε, i.e.,

$$\varepsilon_{ij} = \varepsilon = \varepsilon_i, \qquad i, j = 1, 2, \ldots, n. \qquad (6.41)$$

It will be seen that the formula which we derive resembles closely the corresponding one derived for Method II, the ill-conditioning factor ϕ appearing in the formula for Method IV.

Now, we have by (1.40) that:

When all the $\varepsilon_{ij} = \varepsilon$ then $c_{ki} = b_k \varepsilon,$ $i = 1, 2, \ldots, n.$ (6.42)

Hence, for the case in (6.41) we have by (6.12) that

$$\Delta x_k^{\text{IV}} = \Delta x_k^{\text{II}} + \sum_{i=1}^{n} c_{ki} \, \Delta x_i^{\text{IV}} = \Delta x_k^{\text{II}} + b_k \varepsilon \sum_{i=1}^{n} \Delta x_i^{\text{IV}}, \tag{6.43}$$

$$k = 1, 2, \ldots, n.$$

And by (4.27) we then have

$$\Delta x_k^{\text{IV}} = (1 + x) b_k \varepsilon + b_k \varepsilon \sum_{i=1}^{n} \Delta x_i^{\text{IV}}, \qquad k = 1, 2, \ldots, n. \tag{6.44}$$

Now, writing

$$\delta x \equiv \sum_{i=1}^{n} \Delta x_i^{\text{IV}} = \frac{\text{sum of estimates by Method IV of}}{\text{uncertainties in unknowns,}} \tag{6.45}$$

we have by (6.44) that

$$\Delta x_k^{\text{IV}} = (1 + x) b_k \varepsilon + b_k \varepsilon \, \delta x = (1 + x + \delta x) b_k \varepsilon, \tag{6.46}$$

$$k = 1, 2, \ldots, n.$$

We now express δx in terms of the ill-conditioning factor ϕ. Summing the n equations in (6.46) we have

$$\sum_{k=1}^{n} \Delta x_k^{\text{IV}} = (1 + x + \delta x) \sum_{k=1}^{n} b_k \varepsilon. \tag{6.47}$$

But by (1.41) and (1.36)

$$\sum_{k=1}^{n} b_k \varepsilon = \sum_{k=1}^{n} \sum_{i=1}^{n} |b_{ki}| \varepsilon = \phi. \tag{6.48}$$

Hence, it follows from (6.47) by (6.45) and (6.48) that

$$\delta x = (1 + x + \delta x) \phi, \qquad \text{i.e.,} \qquad (1 - \phi) \, \delta x = (1 + x) \phi. \tag{6.49}$$

Now, in our case the assumption $\rho(\mathbf{C}) < 1$ in (6.20) reduces to

$$\phi < 1 \tag{6.50}$$

because $\rho(\mathbf{C}) = \phi$ when all the $\varepsilon_{ij} = \varepsilon$ (see (2.45)). We therefore have by (6.49) that

$$\delta x = (1 + x) \phi / (1 - \phi), \tag{6.51}$$

this being an expression for δx in terms of ϕ. We now have by (6.46) and (6.51) that

$$\Delta x_k^{\text{IV}} = (1 + x) b_k \varepsilon + (1 + x) b_k \varepsilon \phi / (1 - \phi), \qquad k = 1, 2, \ldots, n,$$

i.e.,

$$\Delta x_k^{IV} = (1 + x)b_k\varepsilon/(1 - \phi), \qquad k = 1, 2, \ldots, n. \qquad (6.52)$$

This then gives the Δx_k^{IV} when the uncertainties in the coefficients and constants are all equal to ε.

And it follows by (4.27) that

$$\Delta x_k^{IV} = \Delta x_k^{II}/(1 - \phi), \qquad k = 1, 2, \ldots, n, \qquad (6.53)$$

or in vector notation

$$\Delta \mathbf{x}^{IV} = \Delta \mathbf{x}^{II}/(1 - \phi). \qquad (6.54)$$

Clearly, if ϕ is small then for the special case being considered the estimates of the uncertainties by Method IV are relatively only slightly greater than those by Method II, the estimates by Method IV being upper bounds of the uncertainties while those by Method II may well be expected to be lower bounds (see (4.66)).

2. Uncertainties in Coefficients all Equal but Constants Exact

We now consider the case

$$\varepsilon_{ij} = \varepsilon, \qquad \varepsilon_i = 0, \qquad i, j = 1, 2, \ldots, n. \qquad (6.55)$$

It is clear by (4.27) and (4.29) that we must replace the first term $(1 + x)b_k\varepsilon$ on the right-hand side of (6.44) by $xb_k\varepsilon$ to obtain the corresponding result for our present case.

Thus we now have

$$\Delta x_k^{IV} = xb_k\varepsilon + b_k\varepsilon \sum_{i=1}^{n} \Delta x_i^{IV}, \qquad k = 1, 2, \ldots, n. \qquad (6.56)$$

Then, carrying out the same steps as those leading to (6.52) we obtain

$$\Delta x_k^{IV} = xb_k\varepsilon/(1 - \phi), \qquad k = 1, 2, \ldots, n. \qquad (6.57)$$

Hence, by (4.29) we again have as in (6.53) that

$$\Delta x_k^{IV} = \Delta x_k^{II}/(1 - \phi), \qquad k = 1, 2, \ldots, n, \qquad (6.58)$$

or in vector notation as in (6.54) that

$$\Delta \mathbf{x}^{IV} = \Delta \mathbf{x}^{II}/(1 - \phi). \qquad (6.59)$$

3. Coefficients Exact but Uncertainties in Constants Unspecified

For the case

$$\varepsilon_{ij} = 0, \qquad \varepsilon_i \text{ unspecified}, \qquad i, j = 1, 2, \ldots, n \qquad (6.60)$$

we have by (6.10) and (4.26) that

$$\Delta x_k^{IV} = \sum_{i=1}^{n} |b_{ki}|\varepsilon_i = \Delta x_k^{II} = \Delta x_k, \qquad k = 1, 2, \ldots, n, \qquad (6.61)$$

or in vector notation

$$\Delta \mathbf{x}^{IV} = |\mathbf{B}|\, \Delta \mathbf{c} = \Delta \mathbf{x}^{II} = \Delta \mathbf{x}. \qquad (6.62)$$

Thus, subject to (6.60) the Δx_k^{IV} give the true uncertainties as do the Δx_k^{II}.

And for the more particular case

$$\varepsilon_{ij} = 0, \qquad \varepsilon_i = \varepsilon, \qquad i, j = 1, 2, \ldots, n,$$

we have by (6.61) that

$$\Delta x_k^{IV} = b_k \varepsilon = \Delta x_k^{II} = \Delta x_k, \qquad k = 1, 2, \ldots, n, \qquad (6.63)$$

or in vector notation

$$\Delta \mathbf{x}^{IV} = \varepsilon \mathbf{b} = \Delta \mathbf{x}^{II} = \Delta \mathbf{x} \qquad (6.64)$$

(see (4.30), (4.31), and (4.34)).

6.5   Uncertainties in Elements of Inverse by Method IV

Let us denote the estimate of the uncertainty by Method IV in the element b_{km} by

$$\Delta b_{km}^{IV}, \qquad k, m = 1, 2, \ldots, n.$$

Then, to obtain the estimates of the uncertainties in the elements in the mth column of the inverse we put

$$\Delta x_k^{IV} = \Delta b_{km}^{IV}, \qquad |x_i| = |b_{im}|, \qquad \varepsilon_i = 0$$

in (6.11) obtaining

$$\Delta b_{km}^{IV} = \sum_{i=1}^{n} c_{ki}(|b_{im}| + \Delta b_{im}^{IV}), \qquad k = 1, 2, \ldots, n. \qquad (6.65)$$

To obtain all the Δb_{km}^{IV} ($k, m = 1, 2, \ldots, n$) we have to solve the system of equations in (6.65) in turn for $m = 1, 2, \ldots, n$.

And corresponding to (6.18) we have

If $\rho(\mathbf{C}) < 1$ then

$$\Delta b_{km}^{IV} \geq \Delta b_{km}, \qquad k, m = 1, 2, \ldots, n. \qquad (6.66)$$

We next determine expressions for the uncertainties in the elements of the inverse by Method IV for the case

$$\varepsilon_{ij} = \varepsilon, \qquad i, j = 1, 2, \ldots, n.$$

By (6.58) and (4.42) we have

$$\Delta b_{km}^{IV} = \Delta b_{km}^{II}/(1 - \phi) = b_k b_{.m} \varepsilon/(1 - \phi), \qquad k, m = 1, 2, \ldots, n, \quad (6.67)$$

or in matrix notation

$$\Delta \mathbf{B}^{IV} = \Delta \mathbf{B}^{II}/(1 - \phi) \tag{6.68}$$

where $\Delta \mathbf{B}^{IV} = (\Delta b_{km}^{IV})$.

And it clearly follows from (6.66) and (6.67) that an upper bound for the uncertainties in the elements of the inverse when all the $\varepsilon_{ij} = \varepsilon$ is given by

$$\|\mathbf{B}\|_{I} \|\mathbf{B}\|_{II} \varepsilon/(1 - \phi)$$

because

$$\Delta b_{km} \leq \Delta b_{km}^{IV} = b_k b_{.m} \varepsilon/(1 - \phi) \leq \|\mathbf{B}\|_{I} \|\mathbf{B}\|_{II} \varepsilon/(1 - \phi),$$
$$k, m = 1, 2, \ldots, n. \tag{6.69}$$

For clearly

$$\|\mathbf{B}\|_{I} = \max_{k}(b_k) \qquad \text{and} \qquad \|\mathbf{B}\|_{II} = \max_{m}(b_{.m})$$

(see also (4.45)).

6.6 Example

To determine the uncertainties in the unknowns by Method IV for the example in (1.43) and (1.44) we use (6.54):

$$\Delta x_k^{IV} = \Delta x_k^{II}/(1 - \phi), \qquad k = 1, 2, \ldots, n.$$

By (4.35) we have

$$\Delta x_1^{II} = 0\cdot1426, \qquad \Delta x_2^{II} = 0\cdot0697, \qquad \Delta x_3^{II} = 0\cdot2193. \quad (6.70)$$

And for $\varepsilon = 0\cdot01$ we have by (1.48) that $\phi = 0\cdot08611$.

Hence

$$\Delta x_1^{IV} = 0\cdot1561, \qquad \Delta x_2^{IV} = 0\cdot0763, \qquad \Delta x_3^{IV} = 0\cdot2400. \quad (6.71)$$

Now, the true uncertainties for the example are given in (10.86):

$$\Delta x_1 = 0\cdot1448, \qquad \Delta x_2 = 0\cdot0710, \qquad \Delta x_3 = 0\cdot2222. \quad (6.72)$$

The estimates by Method IV are clearly upper bounds of the uncertainties, as expected. And these upper bounds are much smaller than the single upper bound of $0\cdot3593$ given by Method I (see (3.25)).

Next, we determine the uncertainties in the elements of the inverse of the coefficient matrix (see (1.45) and (1.46)). These can be obtained using (6.68):

$$\Delta \mathbf{B}^{\text{IV}} = \Delta \mathbf{B}^{\text{II}}/(1 - \phi).$$

Thus, by (4.48) and (1.48) we have

$$\Delta \mathbf{B}^{\text{IV}} = \begin{bmatrix} 0\cdot1122 & 0\cdot0434 & 0\cdot1125 \\ 0\cdot0549 & 0\cdot0212 & 0\cdot0550 \\ 0\cdot1726 & 0\cdot0667 & 0\cdot1730 \end{bmatrix}. \tag{6.73}$$

We note that these upper bounds of the uncertainties in the elements of the inverse are smaller than the upper bounds by Method I (see (3.35)).

And by (6.69) we have

$$\Delta b_{km}^{\text{IV}} \leq \|\mathbf{B}\|_{\text{I}} \|\mathbf{B}\|_{\text{II}} \, \varepsilon/(1 - \phi) = 0\cdot1581/(1 - 0\cdot08611)$$
$$= 0\cdot1730, \qquad k, m = 1, 2, 3 \tag{6.74}$$

(see (4.49)).

This is an upper bound for the largest uncertainty in the elements of the inverse by Method IV and is clearly equal to the largest element in (6.73).

And we may note that the corresponding result of $0\cdot2866$ by Method I in (3.38) is much larger.

6.7   The Method of Solution

By (6.15) the estimates of the uncertainties by Method IV are given by

$$(\mathbf{I} - \mathbf{C}) \, \Delta \mathbf{x}^{\text{IV}} = \Delta \mathbf{x}^{\text{II}} \tag{6.75}$$

where by (4.23)

$$\Delta \mathbf{x}^{\text{II}} = \mathbf{C}|\mathbf{x}| + |\mathbf{B}| \, \Delta \mathbf{c}. \tag{6.76}$$

Now, the solution of (6.75) should cause no difficulty. If a Gaussian elimination scheme was used to solve the original system of equations $\mathbf{A}\mathbf{x} = \mathbf{c}$ it can be used again in (6.75) to solve for the estimates of the uncertainties (see (6.32) and (6.20)).

But the solution of (6.75) by Jacobi iteration or Gauss–Seidel iteration would in practice require less computation. We shall return to these methods in Section 6.9.

Let us now, however, consider the approximate solution of (6.75).

6.8   Approximate Determination of the Δx_k^{IV}

By (6.36) we have

$$\Delta \mathbf{x}^{\text{IV}} = (\mathbf{I} - \mathbf{C})^{-1} \Delta \mathbf{x}^{\text{II}} \qquad \text{if } \rho(\mathbf{C}) < 1. \tag{6.77}$$

But by (2.27) and (2.23)

$$(\mathbf{I} - \mathbf{C})^{-1} = \mathbf{I} + \mathbf{C} + \mathbf{C}^2 + \dots \quad \text{if} \quad \rho(\mathbf{C}) < 1. \quad (6.78)$$

We therefore have by (6.77) and (6.78) that:

$$\text{If} \quad \rho(\mathbf{C}) < 1 \quad \text{then} \quad \Delta\mathbf{x}^{\mathrm{IV}} = (\mathbf{I} + \mathbf{C} + \mathbf{C}^2 + \dots)\,\Delta\mathbf{x}^{\mathrm{II}}. \quad (6.79)$$

And we need not state the condition $\rho(\mathbf{C}) < 1$ explicitly because by (6.20) we can regard it as an implicit assumption when using Method IV.

Now, terminating the series in (6.79) after a finite number of terms leads to an approximate solution of (6.75). And clearly the number of terms to be used depends on the accuracy required and on how large the elements of $\mathbf{C}$ are. If $\|\mathbf{C}\| \ll 1$ for some matrix norm then only two or three terms may be required.

Suppose now that we terminate the series after the first two terms. Then:

$$\text{As a first approximation} \quad \Delta\mathbf{x}^{\mathrm{IV}} \approx (\mathbf{I} + \mathbf{C})\,\Delta\mathbf{x}^{\mathrm{II}}. \quad (6.80)$$

And if we terminate the series after the third term we have:

$$\text{As an improved approximation} \quad \Delta\mathbf{x}^{\mathrm{IV}} \approx (\mathbf{I} + \mathbf{C} + \mathbf{C}^2)\,\Delta\mathbf{x}^{\mathrm{II}}. \quad (6.81)$$

Example

Now, our example in (1.43) and (1.44) falls under the special cases in Section 6.4 and $\Delta\mathbf{x}^{\mathrm{IV}}$ can be easily determined by (6.54). But let us apply the approximation procedures to our example, the condition $\rho(\mathbf{C}) < 1$ being satisfied by (2.49).

We should, however, note that an uncertainty of 0.01 in the coefficients must be regarded as relatively large in comparison to the values of the coefficients. And neither can the elements of the ill-conditioning matrix $\mathbf{C}$ in (1.49) be regarded as small, for

$$\mathbf{C} = \begin{bmatrix} 0.028451 & 0.028451 & 0.028451 \\ 0.013908 & 0.013908 & 0.013908 \\ 0.043755 & 0.043755 & 0.043755 \end{bmatrix}. \quad (6.82)$$

Now, it is easy to check that with $\Delta\mathbf{x}^{\mathrm{II}}$ in (4.35) and $\mathbf{C}$ in (6.82) we have

$$\Delta\mathbf{x}^{\mathrm{II}} = \begin{bmatrix} 0.1426 \\ 0.0697 \\ 0.2193 \end{bmatrix}, \qquad \mathbf{C}\,\Delta\mathbf{x}^{\mathrm{II}} = \begin{bmatrix} 0.012279 \\ 0.006003 \\ 0.018885 \end{bmatrix},$$

$$\mathbf{C}^2\,\Delta\mathbf{x}^{\mathrm{II}} = \begin{bmatrix} 0.001057 \\ 0.000517 \\ 0.001626 \end{bmatrix}, \qquad \mathbf{C}^3\,\Delta\mathbf{x}^{\mathrm{II}} = \begin{bmatrix} 0.000091 \\ 0.000045 \\ 0.000140 \end{bmatrix}. \quad (6.83)$$

Thus, by (6.80) and (6.81) we have the approximate results

$$\Delta \mathbf{x}^{\text{IV}} \approx \begin{bmatrix} 0 \cdot 1549 \\ 0 \cdot 0757 \\ 0 \cdot 2382 \end{bmatrix} \quad \text{and} \quad \Delta \mathbf{x}^{\text{IV}} \approx \begin{bmatrix} 0 \cdot 1559 \\ 0 \cdot 0762 \\ 0 \cdot 2398 \end{bmatrix}, \qquad (6.84)$$

respectively, while by (6.71) we have

$$\Delta \mathbf{x}^{\text{IV}} = \begin{bmatrix} 0 \cdot 1561 \\ 0 \cdot 0763 \\ 0 \cdot 2400 \end{bmatrix}. \qquad (6.85)$$

It is clear that at least one of the approximate values obtained by (6.81) is in error by more than one unit in the fourth decimal place. (Examining $\mathbf{C}^3\,\Delta \mathbf{x}^{\text{II}}$ in (6.83) we see that the error in Δx_3^{IV} by (6.81) is at least 1·4 units in the fourth decimal place.)

And when we examine the way corresponding elements in $\mathbf{C}\,\Delta \mathbf{x}^{\text{II}}$, $\mathbf{C}^2\,\Delta \mathbf{x}^{\text{II}}$, and $\mathbf{C}^3\,\Delta \mathbf{x}^{\text{II}}$ in (6.83) decrease it is clear that the formula

$$\Delta \mathbf{x}^{\text{IV}} \approx (\mathbf{I} + \mathbf{C} + \mathbf{C}^2 + \mathbf{C}^3)\,\Delta \mathbf{x}^{\text{II}} \qquad (6.86)$$

would lead to values of the Δx_k^{IV} to within one unit in the fourth decimal place.

But it is possible to determine directly upper bounds for the errors in the Δx_k^{IV} when using any of the approximations obtained from (6.79). And we now determine an upper bound for the errors when using (6.86).

Upper Bound of Error for Approximation in (6.86)

An upper bound can be obtained by finding the first vector norm of the difference $\Delta \mathbf{x}^{\text{IV}} - (\Delta \mathbf{x}^{\text{IV}})_{\text{approx}}$, the suffix 'approx' implying that the values are obtained by an approximate formula. For we have that

$$|\Delta x_k^{\text{IV}} - (\Delta x^{\text{IV}})_{\text{approx}}| \leq \|\Delta \mathbf{x}^{\text{IV}} - (\Delta \mathbf{x}^{\text{IV}})_{\text{approx}}\|_{\text{I}}, \\ k = 1, 2, \ldots, n, \qquad (6.87)$$

it being recalled that the first vector norm represents the element of maximum magnitude.

Now, in applying (6.87) to the approximation in (6.86) we have

$$\Delta \mathbf{x}^{\text{IV}} = (\mathbf{I} - \mathbf{C})^{-1}\,\Delta \mathbf{x}^{\text{II}} \quad \text{and} \quad (\Delta \mathbf{x}^{\text{IV}})_{\text{approx}} = (\mathbf{I} + \mathbf{C} + \mathbf{C}^2 + \mathbf{C}^3)\,\Delta \mathbf{x}^{\text{II}}$$

(see (6.77)).

Hence,

$$\|\Delta \mathbf{x}^{\text{IV}} - (\Delta \mathbf{x}^{\text{IV}})_{\text{approx}}\|_{\text{I}} = \|((\mathbf{I} - \mathbf{C})^{-1} - (\mathbf{I} + \mathbf{C} + \mathbf{C}^2 + \mathbf{C}^3))\,\Delta \mathbf{x}^{\text{II}}\|_{\text{I}} \\ \leq \|(\mathbf{I} - \mathbf{C})^{-1} - (\mathbf{I} + \mathbf{C} + \mathbf{C}^2 + \mathbf{C}^3)\|_{\text{I}}\,\|\Delta \mathbf{x}^{\text{II}}\|_{\text{I}}.$$

Hence, by (2.31)

$$\|\Delta \mathbf{x}^{IV} - (\Delta \mathbf{x}^{IV})_{\text{approx}}\|_{I} \leq \frac{\|\mathbf{C}\|_{I}^{3+1}}{1 - \|\mathbf{C}\|_{I}} \|\Delta \mathbf{x}^{II}\|_{I} \qquad (6.88)$$

provided

$$\|\mathbf{C}\|_{I} < 1. \qquad (6.89)$$

Now, for our example in (1.43) and (1.44) we have by (2.48) that

$$\|\mathbf{C}\|_{I} = 0 \cdot 1313 \qquad (6.90)$$

so that (6.89) is satisfied. And by (4.35) or (6.83) we have

$$\|\Delta \mathbf{x}^{II}\|_{I} = 0 \cdot 2193. \qquad (6.91)$$

Hence, we have by (6.88) for the example in (1.43) and (1.44) that

$$\|\Delta \mathbf{x}^{IV} - (\Delta \mathbf{x}^{IV})_{\text{approx}}\|_{I} \leq \frac{(0 \cdot 1313)^4}{1 - 0 \cdot 1313} \times 0 \cdot 2193 = 0 \cdot 000075. \quad (6.92)$$

Thus, the approximate formula in (6.86) would lead to errors less than one unit in the fourth decimal place. (We may note that the value of $0 \cdot 000075$ in (6.92) is in fact excessive, this being clear intuitively by considering the values in (6.83), and this conclusion may be checked by comparing corresponding elements in $\Delta \mathbf{x}^{IV}$ and $(\Delta \mathbf{x}^{IV})_{\text{approx}}$ when the elements are computed correct to six decimal places. Norm analysis frequently leads to upper bounds that are far larger than the least upper bound.)

The Number of Terms in Approximate Formula

Let us now ask ourselves whether a computer program can be arranged to determine the number of terms to be taken in (6.79) to obtain a given accuracy.

In particular, let us suppose that the approximate values of the Δx_k^{IV} are required to be correct to within one unit in the dth decimal place so that

$$\|\Delta \mathbf{x}^{IV} - (\Delta \mathbf{x}^{IV})_{\text{approx}}\|_{I} \leq 10^{-d}. \qquad (6.93)$$

If this requirement can be satisfied by taking the terms in (6.79) up to and including $\mathbf{C}^m$, then we have by (6.77) and (6.79) that

$$\Delta \mathbf{x}^{IV} - (\Delta \mathbf{x}^{IV})_{\text{approx}} = ((\mathbf{I} - \mathbf{C})^{-1} - (\mathbf{I} + \mathbf{C} + \mathbf{C}^2 + \ldots + \mathbf{C}^m)) \Delta \mathbf{x}^{II}.$$

Hence, on taking first norms we have by (2.31) that

$$\|\Delta \mathbf{x}^{IV} - (\Delta \mathbf{x}^{IV})_{\text{approx}}\|_{I}$$

$$\leq \|(\mathbf{I} - \mathbf{C})^{-1} - (\mathbf{I} + \mathbf{C} + \mathbf{C}^2 + \ldots + \mathbf{C}^m)\|_{I} \|\Delta \mathbf{x}^{II}\|_{I}$$

$$\leq \frac{\|\mathbf{C}\|_{I}^{m+1}}{1 - \|\mathbf{C}\|_{I}} \|\Delta \mathbf{x}^{II}\|_{I} \qquad \text{provided } \|\mathbf{C}\|_{I} < 1. \qquad (6.94)$$

Hence, provided

$$\|\mathbf{C}\|_I < 1 \tag{6.95}$$

we have by (6.94) and (6.93) that an accuracy of 10^{-d} is achieved if

$$\frac{\|\mathbf{C}\|_I^{m+1}}{1 - \|\mathbf{C}\|_I}\|\Delta \mathbf{x}^{II}\|_I \leq 10^{-d}$$

i.e., if

$$\|\mathbf{C}\|_I^{m+1} \leq (1 - \|\mathbf{C}\|_I)10^{-d}/\|\Delta \mathbf{x}^{II}\|_I. \tag{6.96}$$

Now, taking logarithms to the base 10 we have

$$(m + 1) \log_{10}(\|\mathbf{C}\|_I) \leq \log_{10}(1 - \|\mathbf{C}\|_I) - d - \log_{10}(\|\Delta \mathbf{x}^{II}\|_I).$$

But $\log_{10}(\|\mathbf{C}\|_I) < 0$ by (6.95). Hence,

$$\begin{aligned} m &\geq (\log_{10}(1 - \|\mathbf{C}\|_I) - d - \log_{10}(\|\Delta \mathbf{x}^{II}\|_I))/\log_{10}(\|\mathbf{C}\|_I) - 1 \\ &= L, \text{ say.} \end{aligned} \tag{6.97}$$

Now, m must clearly be an integer because we are taking the terms up to $\mathbf{C}^m$ in (6.79). It follows by (6.97) that the smallest value of m, m_{min}, for (6.96) to hold is given by

$$m_{min} = \text{integer}(L) + 1, \tag{6.98}$$

the *integer function* merely retaining the integer part of a number. (For example, integer (4·967) = 4, integer (−7·238) = −7.)

Let us now determine the value of m_{min} for our example in (1.43) and (1.44) for $d = 4$, i.e., for an error not greater than one unit in the fourth decimal place due to the use of the approximate formula (see (6.93)).

We clearly have by (6.97), (6.90), and (6.91) that

$$\begin{aligned} L &= (\log_{10}(1 - 0\cdot 1313) - 4 - \log_{10}(0\cdot 2193))/\log_{10}(0\cdot 1313) - 1 \\ &= (-0\cdot 0611 - 4 + 0\cdot 6590)/(-0\cdot 8817) - 1 \\ &= 3\cdot 86 - 1 = 2\cdot 86. \end{aligned}$$

Hence, by (6.98)

$$m_{min} = \text{integer}(2\cdot 86) + 1 = 2 + 1 = 3. \tag{6.99}$$

And we have seen in (6.92) that $m = 3$ will indeed achieve the required accuracy.

It may now be noted that L in (6.97) contains the term

$$-d/\log_{10}(\|\mathbf{C}\|_I) \tag{6.100}$$

so that L is a linear function in d. Hence, to determine the number of additional terms to be taken in (6.79) for each additional decimal place of accuracy we put $d = 1$ in $-d/\log_{10}(\|\mathbf{C}\|_1)$. Thus, by (6.90) each additional decimal place of accuracy requires for our example

$$-1/\log_{10}(0 \cdot 1313) = -1/(-0 \cdot 8817) = 1 \cdot 134$$

additional terms in (6.79). And it should be clear that when $\|\mathbf{C}\|_1$ is nearly equal to unity then the approximate procedure outlined in this section may become demanding in computation.

Now, we may note that an efficient method for computing the approximate relations is by the recurrence relation

$$(\mathbf{I} + \mathbf{C} + \mathbf{C}^2 + \ldots + \mathbf{C}^{k+1}) \Delta \mathbf{x}^{\mathrm{II}}$$
$$= \Delta \mathbf{x}^{\mathrm{II}} + \mathbf{C}((\mathbf{I} + \mathbf{C} + \mathbf{C}^2 + \ldots + \mathbf{C}^k) \Delta \mathbf{x}^{\mathrm{II}}),$$

the step on the right-hand side involving the multiplication of the matrix $\mathbf{C}$ and the vector

$$((\mathbf{I} + \mathbf{C} + \mathbf{C}^2 + \ldots + \mathbf{C}^k) \Delta \mathbf{x}^{\mathrm{II}})$$

and the addition of the two vectors

$$\Delta \mathbf{x}^{\mathrm{II}} \quad \text{and} \quad \mathbf{C}((\mathbf{I} + \mathbf{C} + \mathbf{C}^2 + \ldots + \mathbf{C}^k) \Delta \mathbf{x}^{\mathrm{II}}).$$

Thus, the steps to obtain $(\mathbf{I} + \mathbf{C} + \mathbf{C}^2 + \mathbf{C}^3) \Delta \mathbf{x}^{\mathrm{II}}$ in (6.86) would be to compute recurrently

$$\Delta \mathbf{x}^{\mathrm{II}} + \mathbf{C} \Delta \mathbf{x}^{\mathrm{II}} = (\mathbf{I} + \mathbf{C}) \Delta \mathbf{x}^{\mathrm{II}},$$
$$\Delta \mathbf{x}^{\mathrm{II}} + \mathbf{C}((\mathbf{I} + \mathbf{C}) \Delta \mathbf{x}^{\mathrm{II}}) = (\mathbf{I} + \mathbf{C} + \mathbf{C}^2) \Delta \mathbf{x}^{\mathrm{II}},$$

and, finally,

$$\Delta \mathbf{x}^{\mathrm{II}} + \mathbf{C}((\mathbf{I} + \mathbf{C} + \mathbf{C}^2) \Delta \mathbf{x}^{\mathrm{II}}) = (\mathbf{I} + \mathbf{C} + \mathbf{C}^2 + \mathbf{C}^3) \Delta \mathbf{x}^{\mathrm{II}}.$$

In all, then, for $m = 3$, three multiplications of a matrix and a vector and three vector additions are involved.

Should the indicated value of $m_{\min}$ in (6.98) be too large, then clearly an approximation based on (6.79) is not a suitable solution procedure for (6.75).

Next, we consider the solution of (6.75) by iterative methods.

6.9   Solution by Jacobi and Gauss–Seidel Iteration

We now consider the solution of (6.75)

$$(\mathbf{I} - \mathbf{C}) \Delta \mathbf{x}^{\mathrm{IV}} = \Delta \mathbf{x}^{\mathrm{II}} \tag{6.101}$$

by Jacobi and Gauss–Seidel iteration (see reference 1, pp. 56–73).

Generally, these calculation schemes work well if the diagonal elements of the coefficient matrix are far larger in magnitude than the other elements in their respective rows. This situation holds in our case in (6.101), when the elements c_{ij} of the ill-conditioning matrix are small. Further, since the estimates of the uncertainties in the unknowns are only required to a limited accuracy, not more than two or three iterations would be necessary in practice.

Now, Gauss–Seidel iteration is generally faster than Jacobi iteration, because in Gauss–Seidel iteration we use the new value of the iterate as soon as it is obtained. But in our case there is expected to be little difference between the two procedures. For, with the uncertainties ε_{ij} as small as they are in practice so that the elements c_{ij} have small values (unless the elements of the inverse $\mathbf{B}$ are unusually large), it is expected that either procedure would require only a few iteration cycles. Nevertheless, the Gauss–Seidel procedure is slightly easier to program so that it might as well be used.

The number of iteration cycles required depends, of course, on the magnitude of the elements of $\mathbf{C}$ and on the accuracy requirements. But in certain situations the Jacobi and Gauss–Seidel procedures may not converge.

 a well-known sufficient condition for the
tive processes for an arbitrary initial guess.
at the coefficient matrix be strictly diagonally
cient matrix $\mathbf{H} = (h_{ij})$ it requires that

$$|h_{ij}|, \qquad i = 1, 2, \ldots, n. \tag{6.102}$$

this sufficient condition for convergence re-

$$> \sum_{\substack{j \neq i \\ j=1}}^{n} c_{ij}, \qquad i = 1, 2, \ldots, n, \tag{6.103}$$

a nonnegative matrix.

And we now show that (6.103) is satisfied if

$$\|\mathbf{C}\|_{\mathrm{I}} < 1 \tag{6.104}$$

where

$$\|\mathbf{C}\|_{\mathrm{I}} = \max_i \left(\sum_{j=1}^{n} c_{ij} \right),$$

the elements of $\mathbf{C}$ being nonnegative.

For if $\|\mathbf{C}\|_{\mathrm{I}} < 1$ then

$$\sum_{j=1}^{n} c_{ij} < 1, \qquad i = 1, 2, \ldots, n,$$

so that

$$1 - c_{ii} > \sum_{\substack{j \neq i \\ j=1}}^{n} c_{ij}, \qquad i = 1, 2, \ldots, n. \tag{6.105}$$

The requirement in (6.103) is then clearly satisfied.

Thus, (6.104) is a sufficient condition for the convergence of the Jacobi and Gauss–Seidel procedures for the determination of $\Delta \mathbf{x}^{IV}$ by (6.101).

We note, however, that (6.104) does not cover our complete field of interest as specified by $\rho(\mathbf{C}) < 1$ (see (6.20) and (2.24)). But the Jacobi or Gauss–Seidel iteration processes are used in practice usually only when the coefficient matrix is strongly diagonally dominant, i.e., when $\|\mathbf{C}\|_{\mathrm{I}} \ll 1$ in (6.101). Thus, the condition in (6.104) is satisfactory.

Similarly, the approximate method in Section 6.8 is also only indicated when $\|\mathbf{C}\|_{\mathrm{I}}$ is small (see (6.88) and (6.89)).

And, of course, the solution by Gaussian elimination is possible in all cases where $\rho(\mathbf{C}) < 1$ and is especially indicated when the other methods are not suitable (see (6.36)).

In conclusion, we may note that we shall give an alternative proof in Section 9.7 to the one given in Section 6.3, showing that the Δx_k^{IV} are upper bounds of the uncertainties provided $\rho(\mathbf{C}) < 1$.

Reference

1. VARGA, R. S. *Matrix Iterative Analysis*, Prentice-Hall, Englewood Cliffs, N.J. (1962).

CHAPTER 7

Method V: Lower Bounds of Uncertainties

7.1 Introduction

We have seen in (6.18) that Δx_k^{IV} is an upper bound of the uncertainty Δx_k, while by (4.66) Δx_k^{II} is a lower bound of Δx_k at a distance of

$$\Delta x_k^{IV} - \Delta x_k^{II} \tag{7.1}$$

from Δx_k^{IV}, subject to certain assumptions being valid, and this holds for $k = 1, 2, \ldots, n$.

But it will be clear from the proof in the next section that if

$$\rho(\mathbf{C}) < 1 \tag{7.2}$$

then lower bounds of the Δx_k certainly lie at distances

$$2(\Delta x_k^{IV} - \Delta x_k^{II}), \qquad k = 1, 2, \ldots, n, \tag{7.3}$$

from the upper bounds Δx_k^{IV}, i.e.,

$$\begin{aligned}
\Delta x_k \in [\Delta x_k^{IV} &- 2(\Delta x_k^{IV} - \Delta x_k^{II}), \Delta x_k^{IV}] \\
&= [2\Delta x_k^{II} - \Delta x_k^{IV}, \Delta x_k^{IV}], \qquad k = 1, 2, \ldots, n,
\end{aligned} \tag{7.4}$$

subject to (7.2).

We now define the estimates by Method V to be the lower endpoints of the intervals in (7.4):

$$\Delta x_k^{V} \equiv 2\Delta x_k^{II} - \Delta x_k^{IV}, \qquad k = 1, 2, \ldots, n, \tag{7.5}$$

i.e.,

$$\Delta \mathbf{x}^{V} \equiv 2\Delta \mathbf{x}^{II} - \Delta \mathbf{x}^{IV} \tag{7.6}$$

where $\Delta \mathbf{x}^{V} = (\Delta x_k^{V})$.

In the next section we prove that:

$$\text{If} \quad \rho(\mathbf{C}) < 1 \quad \text{then} \quad \Delta \mathbf{x}^{V} \le \Delta \mathbf{x}. \tag{7.7}$$

It then follows by (6.18) that:

$$\text{If } \rho(\mathbf{C}) < 1 \quad \text{then} \quad \Delta\mathbf{x} \in [\Delta\mathbf{x}^{\text{V}}, \Delta\mathbf{x}^{\text{IV}}]. \tag{7.8}$$

And the results in (7.6), (7.7), and (7.8) include those in (7.2), (7.3), and (7.4).

7.2   Proof that the Δx_k^{V} are Lower Bounds of the Uncertainties

We now prove the result in (7.7):

$$\text{If } \rho(\mathbf{C}) < 1 \quad \text{then} \quad \Delta\mathbf{x}^{\text{V}} \le \Delta\mathbf{x}. \tag{7.9}$$

Let us then assume that

$$\rho(\mathbf{C}) < 1 \tag{7.10}$$

(see (6.20) and (2.36)).

Now, let us consider the δa_{ij} and δc_i in (4.51) for some fixed k. Then by (4.58) we have

$$d_{kk} = \Delta x_k^{\text{II}} + \sum_{j=1}^{n} c_{kj}d_{jk}\,\text{sign}(x_j) = \Delta x_k^{\text{II}} + \delta_k \tag{7.11}$$

where

$$\delta_k = \sum_{j=1}^{n} c_{kj}d_{jk}\,\text{sign}(x_j). \tag{7.12}$$

But by (4.52) the d_{jk} $(j = 1, 2, \ldots, n)$ are changes in the x_j $(j = 1, 2, \ldots, n)$ for changes in the coefficients and constants within the limits of their uncertainties and by (7.10) and (2.36) the uncertainties Δx_j exist. Hence, we have that

$$|d_{jk}| \le \Delta x_j, \quad j = 1, 2, \ldots, n, \tag{7.13}$$

and in particular that

$$|d_{kk}| \le \Delta x_k. \tag{7.14}$$

Now, taking moduli in (7.12) we have

$$|\delta_k| \le \sum_{j=1}^{n} c_{kj}\,|d_{jk}|\,|\text{sign}(x_j)|$$

$$\le \sum_{j=1}^{n} c_{kj}\,\Delta x_j, \quad \text{by (7.13)},$$

$$\le \sum_{j=1}^{n} c_{kj}\,\Delta x_j^{\text{IV}}, \quad \text{by (6.18) and (7.10)},$$

$$= \Delta x_k^{\text{IV}} - \Delta x_k^{\text{II}}, \quad \text{by (6.12)}.$$

Thus

$$|\delta_k| \leq \Delta x_k^{\mathrm{IV}} - \Delta x_k^{\mathrm{II}}. \tag{7.15}$$

But

$$\Delta x_k^{\mathrm{II}} = d_{kk} - \delta_k \tag{7.16}$$

by (7.11) so that on taking moduli we have

$$\Delta x_k^{\mathrm{II}} \leq |d_{kk}| + |\delta_k|$$
$$\leq \Delta x_k + \Delta x_k^{\mathrm{IV}} - \Delta x_k^{\mathrm{II}}, \qquad \text{by (7.14) and (7.15).}$$

Thus

$$2\Delta x_k^{\mathrm{II}} - \Delta x_k^{\mathrm{IV}} \leq \Delta x_k,$$

i.e.,

$$\Delta x_k^{\mathrm{V}} \leq \Delta x_k \tag{7.17}$$

(see (7.5)).

And this result must hold for $k = 1, 2, \ldots, n$. Hence we have by (7.10) and (7.17) that

$$\Delta \mathbf{x}^{\mathrm{V}} \leq \Delta \mathbf{x} \qquad \text{if} \qquad \rho(\mathbf{C}) < 1. \tag{7.18}$$

This, then, is the result we set out to prove.

Hence, by (6.18) and (7.18) the uncertainties Δx_k can be bounded both above and below, the intervals being given in (7.4) and (7.8).

In using the Δx_k^{V}, the assumption $\rho(\mathbf{C}) < 1$ need not be stated explicitly everywhere, because it can be regarded as an implicit assumption of the method (see (7.10)). (And by (6.20) this is also true of Method IV.)

Intervals Bounding the Uncertainties

In the rest of this section, we express the interval in (7.8) in various convenient forms.

First we note that $\Delta \mathbf{x}^{\mathrm{V}}$ can be expressed in terms of $\Delta \mathbf{x}^{\mathrm{IV}}$ and $\mathbf{C}$ only. For by (6.15) and (7.6) we have

$$\Delta \mathbf{x}^{\mathrm{V}} = 2\Delta \mathbf{x}^{\mathrm{II}} - \Delta \mathbf{x}^{\mathrm{IV}} = 2(\mathbf{I} - \mathbf{C})\,\Delta \mathbf{x}^{\mathrm{IV}} - \Delta \mathbf{x}^{\mathrm{IV}}$$
$$= (\mathbf{I} - 2\mathbf{C})\,\Delta \mathbf{x}^{\mathrm{IV}}. \tag{7.19}$$

Hence, by (7.8) and (7.19) we have

$$\Delta \mathbf{x} \in [(\mathbf{I} - 2\mathbf{C})\,\Delta \mathbf{x}^{\mathrm{IV}},\, \Delta \mathbf{x}^{\mathrm{IV}}]. \tag{7.20}$$

Now, let us note by (7.6) and (7.8) that

$$\Delta \mathbf{x} \in [\Delta \mathbf{x}^{\mathrm{II}} - (\Delta \mathbf{x}^{\mathrm{IV}} - \Delta \mathbf{x}^{\mathrm{II}}),\, \Delta \mathbf{x}^{\mathrm{II}} + (\Delta \mathbf{x}^{\mathrm{IV}} - \Delta \mathbf{x}^{\mathrm{II}})], \tag{7.21}$$

i.e., $\Delta \mathbf{x}^{\mathrm{II}}$ is the midpoint of the vector interval in (7.8).

But by (6.15) we have

$$\Delta \mathbf{x}^{IV} - \Delta \mathbf{x}^{II} = \Delta \mathbf{x}^{IV} - (\mathbf{I} - \mathbf{C}) \Delta \mathbf{x}^{IV} = \mathbf{C} \Delta \mathbf{x}^{IV}. \qquad (7.22)$$

Hence, by (7.21) and (7.22) we have that

$$\Delta \mathbf{x} \in [\Delta \mathbf{x}^{II} - \mathbf{C} \Delta \mathbf{x}^{IV}, \Delta \mathbf{x}^{II} + \mathbf{C} \Delta \mathbf{x}^{IV}]. \qquad (7.23)$$

The result in (7.8) may thus be stated as follows:

> If $\rho(\mathbf{C}) < 1$ then the vector interval with midpoint $\Delta \mathbf{x}^{II}$ and half-width $\mathbf{C} \Delta \mathbf{x}^{IV}$ bounds the uncertainty vector $\Delta \mathbf{x}$. $\qquad (7.24)$

Now, by (6.16) we have

$$\mathbf{C} \Delta \mathbf{x}^{IV} = \mathbf{C}(\mathbf{I} - \mathbf{C})^{-1} \Delta \mathbf{x}^{II} = \mathbf{C}(\mathbf{I} + \mathbf{C} + \mathbf{C}^2 + \ldots) \Delta \mathbf{x}^{II} \qquad (7.25)$$

if $\rho(\mathbf{C}) < 1$.

Hence, it follows by (7.23) and (7.25) that:
If $\rho(\mathbf{C}) < 1$ then

$$\Delta \mathbf{x} \in [(\mathbf{I} - \mathbf{C} - \mathbf{C}^2 - \ldots) \Delta \mathbf{x}^{II}, (\mathbf{I} + \mathbf{C} + \mathbf{C}^2 + \ldots) \Delta \mathbf{x}^{II}], \qquad (7.26)$$

a form suitable when considering the errors in the approximate determination of the intervals bounding the uncertainties.

Thus, other forms of the interval in (7.8) bounding the uncertainty vector above and below are given in (7.20), (7.21), (7.23), and (7.26).

7.3   Special Cases

1. Uncertainties in Coefficients and Constants all Equal

When

$$\varepsilon_{ij} = \varepsilon = \varepsilon_i, \qquad i, j = 1, 2, \ldots, n, \qquad (7.27)$$

we have by (6.54) that

$$\Delta \mathbf{x}^{IV} = \Delta \mathbf{x}^{II}/(1 - \phi), \qquad \text{i.e.,} \qquad \Delta \mathbf{x}^{II} = (1 - \phi) \Delta \mathbf{x}^{IV}. \qquad (7.28)$$

Hence, by (7.6)

$$\Delta \mathbf{x}^{V} = 2\Delta \mathbf{x}^{II} - \Delta \mathbf{x}^{IV} = 2(1 - \phi) \Delta \mathbf{x}^{IV} - \Delta \mathbf{x}^{IV}$$
$$= (1 - 2\phi) \Delta \mathbf{x}^{IV}. \qquad (7.29)$$

Hence, when the uncertainties in the coefficients and constants are all equal to ε we have by (7.8) and (7.29) that

$$\Delta \mathbf{x} \in [\Delta \mathbf{x}^{V}, \Delta \mathbf{x}^{IV}] = [(1 - 2\phi) \Delta \mathbf{x}^{IV}, \Delta \mathbf{x}^{IV}], \qquad (7.30)$$

where by (7.28) and (4.32)

$$\Delta \mathbf{x}^{IV} = \Delta \mathbf{x}^{II}/(1 - \phi) \qquad \text{with} \qquad \Delta \mathbf{x}^{II} = (1 + x)\varepsilon \mathbf{b}. \qquad (7.31)$$

And we can express the intervals in (7.30) directly in terms of $\Delta \mathbf{x}^{II}$:

$$\Delta \mathbf{x} \in [(1 - 2\phi) \Delta \mathbf{x}^{II}/(1 - \phi), \Delta \mathbf{x}^{II}/(1 - \phi)]. \qquad (7.32)$$

It is clear by (7.30) and (7.32) that the intervals bounding the Δx_k are small if the ill-conditioning factor is small, provided the Δx_k^{II} or Δx_k^{IV} are not too large.

2. Uncertainties in Coefficients all Equal but Constants Exact

When

$$\varepsilon_{ij} = \varepsilon, \qquad \varepsilon_i = 0, \qquad i, j = 1, 2, \ldots, n, \qquad (7.33)$$

we have by (6.59) that

$$\Delta \mathbf{x}^{IV} = \Delta \mathbf{x}^{II}/(1 - \phi), \qquad \text{i.e.,} \qquad \Delta \mathbf{x}^{II} = (1 - \phi) \Delta \mathbf{x}^{IV}. \quad (7.34)$$

And in our present case

$$\Delta \mathbf{x}^{II} = x \varepsilon \mathbf{b}, \qquad \Delta \mathbf{x}^{IV} = x \varepsilon \mathbf{b}/(1 - \phi), \qquad (7.35)$$

by (4.33) and (7.34).

Now, the equations in (7.28) and (7.34) are of the same form. Hence:

The results in (7.29), (7.30), and (7.32) hold when the uncertainties in the coefficients are all equal but the constants are exact, $\Delta \mathbf{x}^{II}$ and $\Delta \mathbf{x}^{IV}$ being now given by (7.35).   (7.36)

3. Coefficients Exact but Uncertainties in Constants Unspecified

For the case

$$\Delta \mathbf{A} = (\varepsilon_{ij}) = \mathbf{0}, \qquad \Delta \mathbf{c} = (\varepsilon_i) \text{ unspecified} \qquad (7.37)$$

we have by (6.62) that

$$\Delta \mathbf{x}^{IV} = |\mathbf{B}| \Delta \mathbf{c} = \Delta \mathbf{x}^{II} = \Delta \mathbf{x}. \qquad (7.38)$$

Hence, by (7.5)

$$\Delta \mathbf{x}^{V} = \Delta \mathbf{x}. \qquad (7.39)$$

Including this result in (7.38) we have:

$\Delta \mathbf{x} = \Delta \mathbf{x}^{V} = \Delta \mathbf{x}^{IV} = \Delta \mathbf{x}^{II} = |\mathbf{B}| \Delta \mathbf{c}$ when uncertainties exist only in the right-hand constants, the coefficients being known exactly.   (7.40)

And for the more particular case

$$\varepsilon_{ij} = 0, \qquad \varepsilon_i = \varepsilon, \qquad i, j = 1, 2, \ldots, n, \qquad (7.41)$$

we then have

$$\Delta \mathbf{x} = \Delta \mathbf{x}^{\mathrm{V}} = \Delta \mathbf{x}^{\mathrm{IV}} = \Delta \mathbf{x}^{\mathrm{II}} = \varepsilon \mathbf{b} = \varepsilon(b_k) \tag{7.42}$$

(see (4.30), (4.31), and (4.34)).

7.4   Uncertainties in Elements of Inverse by Method V

We denote the estimate of the uncertainty by Method V in the element b_{km} by

$$\Delta b^{\mathrm{V}}_{km}, \qquad k, m = 1, 2, \ldots, n, \tag{7.43}$$

and in vector notation we write

$$\Delta \mathbf{B}^{\mathrm{V}} = (\Delta b^{\mathrm{V}}_{km}).$$

Thus, by (7.6) and (7.7) we have

$$\Delta \mathbf{B}^{\mathrm{V}} = 2\Delta \mathbf{B}^{\mathrm{II}} - \Delta \mathbf{B}^{\mathrm{IV}} \leq \Delta \mathbf{B} \tag{7.44}$$

provided $\rho(\mathbf{C}) < 1$.

And by (7.8) we have

$$\Delta \mathbf{B} \in [\Delta \mathbf{B}^{\mathrm{V}}, \Delta \mathbf{B}^{\mathrm{IV}}]. \tag{7.45}$$

Now for the case

$$\varepsilon_{ij} = \varepsilon, \qquad i, j = 1, 2, \ldots, n, \tag{7.46}$$

we have by (7.36), (7.30), and (6.67) that

$$\Delta \mathbf{B} \in [(1 - 2\phi)\, \Delta \mathbf{B}^{\mathrm{IV}}, \Delta \mathbf{B}^{\mathrm{IV}}] \tag{7.47}$$

in which $\Delta \mathbf{B}^{\mathrm{IV}} = (\Delta b^{\mathrm{IV}}_{km})$ is given by

$$\Delta b^{\mathrm{IV}}_{km} = \Delta b^{\mathrm{II}}_{km}/(1 - \phi) = b_k b_{\cdot m}\varepsilon/(1 - \phi), \quad k, m = 1, 2, \ldots, n. \tag{7.48}$$

We thus have by (7.47) and (7.48) that:

Length of interval bounding $\Delta b_{km} = 2\phi b_k b_{\cdot m}\varepsilon/(1 - \phi),$
$$k, m = 1, 2, \ldots, n. \tag{7.49}$$

Clearly, the lengths of the intervals bounding the uncertainties are small if the elements of the inverse are not large and if ε is small. For then ϕ is small and b_k and $b_{\cdot m}$ are not large.

7.5   Example

To determine the uncertainties in the unknowns by Method V for the example in (1.43) and (1.44) we have by (7.29) that

$$\Delta x^{\mathrm{V}}_k = (1 - 2\phi)\, \Delta x^{\mathrm{IV}}_k, \qquad k = 1, 2, 3. \tag{7.50}$$

But by (1.48) $\phi = 0.08611$. Hence by (6.71) and (7.50) we have

$$\Delta x_1^V = 0.1292, \qquad \Delta x_2^V = 0.0631, \qquad \Delta x_3^V = 0.1986. \qquad (7.51)$$

Thus, by (7.30), (6.71), and (7.51) we have

$$\begin{aligned}
\Delta x_1 &\in [0.1292, 0.1561] \\
\Delta x_2 &\in [0.0631, 0.0763]. \\
\Delta x_3 &\in [0.1986, 0.2400]
\end{aligned} \qquad (7.52)$$

These, then, are intervals bounding the uncertainties in the unknowns above and below, the estimates by Method IV being upper bounds of the uncertainties, while those by Method V are lower bounds.

7.6　The Method of Solution

By (7.19)

$$\Delta \mathbf{x}^V = (\mathbf{I} - 2\mathbf{C})\,\Delta \mathbf{x}^{IV} \qquad (7.53)$$

so that the problem of computing $\Delta \mathbf{x}^V$ is essentially that of computing $\Delta \mathbf{x}^{IV}$. And the computation of $\Delta \mathbf{x}^{IV}$ has been discussed in some detail in Sections 6.7 to 6.9.

It is indeed convenient that the one uncertainty vector can be easily computed from the other. For both $\Delta \mathbf{x}^{IV}$ and $\Delta \mathbf{x}^V$ appear in the interval bounding $\Delta \mathbf{x}$ in (7.8)

$$[\Delta \mathbf{x}^V, \Delta \mathbf{x}^{IV}]. \qquad (7.54)$$

Approximate Determination of Interval Bounding Δx

We have seen in Section 6.8 that $\Delta \mathbf{x}^{IV}$ can be computed approximately by terminating the series in (6.79), error bounds being given by (6.94). And (6.97) and (6.98) give the number of terms to be included for a specified accuracy.

Now by (7.23), (7.25), and (7.26) the lower endvector (i.e., the vector at the lower end) in (7.54) is given by

$$\Delta \mathbf{x}^V = (\mathbf{I} - \mathbf{C} - \mathbf{C}^2 - \ldots)\,\Delta \mathbf{x}^{II} \qquad (7.55)$$

while the upper endvector in (7.54) is

$$\Delta \mathbf{x}^{IV} = (\mathbf{I} + \mathbf{C} + \mathbf{C}^2 + \ldots)\,\Delta \mathbf{x}^{II}. \qquad (7.56)$$

Suppose, now, that we terminate each series at the term $\mathbf{C}^m$. Then, by (7.55) and (7.56) it follows that the error in $\Delta \mathbf{x}^V$ due to the approximation will be equal and opposite to that in $\Delta \mathbf{x}^{IV}$, i.e.,

$$(\Delta \mathbf{x}^V)_{\text{approx}} - \Delta \mathbf{x}^V = \Delta \mathbf{x}^{IV} - (\Delta \mathbf{x}^{IV})_{\text{approx}}, \qquad (7.57)$$

the value of m to be taken for a given accuracy and the error bound being given by (6.98) and (6.94), respectively.

And let us note that having computed $(\Delta\mathbf{x}^{IV})_{approx}$ as indicated at the end of Section 6.8 then we obtain $(\Delta\mathbf{x}^{V})_{approx}$ from

$$(\Delta\mathbf{x}^{V})_{approx} = 2\Delta\mathbf{x}^{II} - (\Delta\mathbf{x}^{IV})_{approx}. \qquad (7.58)$$

For on terminating (7.55) and (7.56) at the same term we clearly have

$$(\Delta\mathbf{x}^{V})_{approx} + (\Delta\mathbf{x}^{IV})_{approx} = 2\Delta\mathbf{x}^{II}. \qquad (7.59)$$

For $m = 3$ we obtain the following approximation by (7.54), (7.55), and (7.56) for the interval bounding $\Delta\mathbf{x}$:

$$[(\mathbf{I} - \mathbf{C} - \mathbf{C}^2 - \mathbf{C}^3)\,\Delta\mathbf{x}^{II}, (\mathbf{I} + \mathbf{C} + \mathbf{C}^2 + \mathbf{C}^3)\,\Delta\mathbf{x}^{II}]. \qquad (7.60)$$

And by (6.99) this leads to results for our example in (1.43) and (1.44) which do not differ by more than one unit in the fourth decimal place from those in (7.52).

Thus, computing $\Delta\mathbf{x}^{V}$ is merely an extension of the computation of $\Delta\mathbf{x}^{IV}$, whether the exact relation in (7.53) is used or the approximate relations obtained by terminating the series in (7.55) and (7.56).

7.7 Further Consideration of Method II

Bounds for Uncertainties in terms of the Δx_i^{II} and the x_i

We now show that it is possible to obtain satisfactory bounds for the uncertainties using only the values of the Δx_i^{II} and the unknowns x_i when

$$\Delta x_i^{II}/|x_i| \ll 1, \qquad i = 1, 2, \ldots, n. \qquad (7.61)$$

The present approach covers the situation in which the Δx_i^{II} and the x_i have been obtained and the bounds for the uncertainties Δx_i are required as an afterthought. For, otherwise, bounds could have been obtained by finding $\Delta\mathbf{x}^{IV}$ and $\Delta\mathbf{x}^{V}$.

First, we derive preliminary results in (7.70) and (7.72) and then bounds for the uncertainties in (7.75).

We now write (6.11) in the form

$$\Delta x_k^{IV} = \sum_{i=1}^{n} c_{ki}\,|x_i|(1 + \Delta x_i^{IV}/|x_i|) + \sum_{i=1}^{n} |b_{ki}|\varepsilon_i, \qquad (7.62)$$

$$k = 1, 2, \ldots, n,$$

assuming that

$$x_i \neq 0, \qquad i = 1, 2, \ldots, n. \qquad (7.63)$$

And by (4.22)

$$\Delta x_k^{\mathrm{II}} = \sum_{i=1}^{n} c_{ki} |x_i| + \sum_{i=1}^{n} |b_{ki}|\varepsilon_i, \qquad k = 1, 2, \ldots, n. \qquad (7.64)$$

Hence

$$\frac{\Delta x_k^{\mathrm{IV}}}{\Delta x_k^{\mathrm{II}}} = \frac{\displaystyle\sum_{i=1}^{n} c_{ki} |x_i|(1 + \Delta x_i^{\mathrm{IV}}/|x_i|) + \sum_{i=1}^{n} |b_{ki}|\varepsilon_i}{\displaystyle\sum_{i=1}^{n} c_{ki} |x_i| + \sum_{i=1}^{n} |b_{ki}|\varepsilon_i}, \qquad k = 1, 2, \ldots, n,$$

$$\leq \frac{\displaystyle\sum_{i=1}^{n} c_{ki} |x_i|(1 + \Delta x_i^{\mathrm{IV}}/|x_i|)}{\displaystyle\sum_{i=1}^{n} c_{ki} |x_i|}, \qquad \text{on subtracting the term } \sum_{i=1}^{n} |b_{ki}|\varepsilon_i \text{ from both the numerator and denominator,}$$

$$\leq 1 + \max_i\left(\frac{\Delta x_i^{\mathrm{IV}}}{|x_i|}\right),$$

since the ith term in the numerator is a factor $(1 + \Delta x_i^{\mathrm{IV}}/|x_i|)$ of the corresponding term in the denominator.

Thus,

$$\frac{\Delta x_k^{\mathrm{IV}}}{\Delta x_k^{\mathrm{II}}} \leq 1 + d, \qquad k = 1, 2, \ldots, n, \qquad (7.65)$$

where

$$d = \max_i(\Delta x_i^{\mathrm{IV}}/|x_i|). \qquad (7.66)$$

But in our present approach the Δx_i^{IV} are not available while the Δx_i^{II} and the x_i are available.

Writing, then,

$$e = \max_i(\Delta x_i^{\mathrm{II}}/|x_i|) \qquad (7.67)$$

we have

$$d = \max_i\left(\frac{\Delta x_i^{\mathrm{IV}}}{|x_i|}\right) = \max_i\left(\frac{\Delta x_i^{\mathrm{II}}}{|x_i|} \frac{\Delta x_i^{\mathrm{IV}}}{\Delta x_i^{\mathrm{II}}}\right)$$

$$\leq \max_i\left(\frac{\Delta x_i^{\mathrm{II}}}{|x_i|}\right) \max_i\left(\frac{\Delta x_i^{\mathrm{IV}}}{\Delta x_i^{\mathrm{II}}}\right) \leq e(1 + d)$$

by (7.67) and (7.65).

Thus $d - ed \leq e$ so that

$$d \leq e/(1 - e), \qquad (7.68)$$

assuming that

$$e < 1. \tag{7.69}$$

Hence, by (7.65) and (7.68) we have

$$\Delta x_k^{IV}/\Delta x_k^{II} \le 1 + d \le 1 + e/(1 - e) = 1/(1 - e), \quad k = 1, 2, \ldots, n,$$

so that

$$\Delta x_k^{IV} \le \Delta x_k^{II}/(1 - e), \quad k = 1, 2, \ldots, n, \tag{7.70}$$

or in vector notation

$$\Delta \mathbf{x}^{IV} \le (1/(1 - e)) \, \Delta \mathbf{x}^{II}. \tag{7.71}$$

And by (7.5) and (7.70) we have

$$\Delta x_k^{V} = 2\Delta x_k^{II} - \Delta x_k^{IV} \ge 2\Delta x_k^{II} - \Delta x_k^{II}/(1 - e)$$
$$= ((1 - 2e)/(1 - e)) \, \Delta x_k^{II},$$

i.e.,

$$\Delta x_k^{V} \ge ((1 - 2e)/(1 - e)) \, \Delta x_k^{II}, \quad k = 1, 2, \ldots, n, \tag{7.72}$$

or in vector notation

$$\Delta \mathbf{x}^{V} \ge ((1 - 2e)/(1 - e)) \, \Delta \mathbf{x}^{II}. \tag{7.73}$$

Hence, by (7.8), (7.71), and (7.73)

$$\Delta \mathbf{x} \in [\Delta \mathbf{x}^{V}, \Delta \mathbf{x}^{IV}] \subseteq [((1 - 2e)/(1 - e)) \, \Delta \mathbf{x}^{II}, (1/(1 - e)) \, \Delta \mathbf{x}^{II}]. \tag{7.74}$$

Thus

$$\Delta \mathbf{x} \in [((1 - 2e)/(1 - e)) \, \Delta \mathbf{x}^{II}, (1/(1 - e)) \, \Delta \mathbf{x}^{II}] \tag{7.75}$$

so that the uncertainty vector $\Delta \mathbf{x}$ is bounded above and below in terms of $\Delta \mathbf{x}^{II}$ and $e = \max_i(\Delta x_i^{II}/|x_i|)$.

Now, when e is sufficiently small the bounds for $\Delta \mathbf{x}$ given by (7.75) are satisfactory in practice. But assumptions were made in (7.63) and (7.69). And by (7.74) the uncertainty vector $\Delta \mathbf{x}$ can only be less closely bounded by (7.75) than by (7.8).

Example

For our numerical example in (1.43) and (1.44) we have by (4.37), (1.47), and (7.67) that

$$e = 0{\cdot}1426/0{\cdot}9730 = 0{\cdot}1466. \tag{7.76}$$

Hence by (7.75), (4.37), and (7.76) we have

$$\begin{aligned}
\Delta x_1 &\in [0{\cdot}1181, \quad 0{\cdot}1671] \\
\Delta x_2 &\in [0{\cdot}0577, \quad 0{\cdot}0817]. \\
\Delta x_3 &\in [0{\cdot}1816, \quad 0{\cdot}2570]
\end{aligned} \tag{7.77}$$

On comparing the intervals in (7.77) with those in (7.52), it is clear that the uncertainties are less closely bounded by (7.75) than by (7.8).

The Relative Errors in the Δx_k^{II}

We now determine an upper bound for the relative errors in the Δx_k^{II} in their representation of the Δx_k in terms of e.

First, we note that $\Delta \mathbf{x}^{II}$ is the midpoint of the vector interval in (7.75), half the sum of the endvectors in (7.75) being equal to $\Delta \mathbf{x}^{II}$. Hence $|\Delta \mathbf{x}^{II} - \Delta \mathbf{x}|$ is not greater than half the difference between the endvectors in (7.75), i.e.,

$$|\Delta \mathbf{x}^{II} - \Delta \mathbf{x}| \leq \tfrac{1}{2}(1/(1 - e) - (1 - 2e)/(1 - e)) \Delta \mathbf{x}^{II}$$
$$= ((e/(1 - e)) \Delta \mathbf{x}^{II}. \qquad (7.78)$$

But considering the lower endvector in (7.75) we have that

$$((1 - 2e)/(1 - e)) \Delta \mathbf{x}^{II} \leq \Delta \mathbf{x}$$

so that

$$\Delta \mathbf{x}^{II} \leq ((1 - e)/(1 - 2e)) \Delta \mathbf{x} \qquad (7.79)$$

provided

$$e < \tfrac{1}{2}. \qquad (7.80)$$

Hence, by (7.78) and (7.79) we have

$$|\Delta \mathbf{x}^{II} - \Delta \mathbf{x}| \leq ((e/(1 - 2e)) \Delta \mathbf{x}. \qquad (7.81)$$

It follows that

$$|\Delta x_k^{II} - \Delta x_k|/\Delta x_k \leq e/(1 - 2e), \qquad k = 1, 2, \ldots, n. \qquad (7.82)$$

Thus, the relative errors in the Δx_k^{II} in representing the Δx_k are certainly small when e is small, i.e., when the Δx_k^{II} are small compared with the absolute values of the corresponding unknowns. In situations in which $\Delta \mathbf{x}^{II}$ alone is determined the magnitude of the ratio $e/(1 - 2e)$ in (7.82) can be used as a criterion to decide whether $\Delta \mathbf{x}^{IV}$ and $\Delta \mathbf{x}^V$ should also be determined.

We have thus shown in this section that useful information concerning how close the Δx_k^{II} are to the Δx_k can be obtained from the ratios $\Delta x_k^{II}/|x_k|$.

CHAPTER 8

Method VI: Upper Bounds of Uncertainties by Combining Methods III and IV

8.1 Introduction

We recall that if the assumption in Method III (see (5.1)) that the partial derivatives

$$\frac{\partial x_k^*}{\partial a_{ij}^*} = -x_j^* b_{ki}^* \quad \text{and} \quad \frac{\partial x_k^*}{\partial c_i^*} = b_{ki}^*, \qquad i, j, k = 1, 2, \ldots, n, \qquad (8.1)$$

do not change sign within the limits of the uncertainties in the coefficients and constants is valid then Method III gives the true intervals of uncertainty, i.e.,

$$U_k^{\mathrm{III}} = U_k, \qquad k = 1, 2, \ldots, n.$$

And we recall further that to determine the maximum increase in x_k by Method III we introduce the following changes in the coefficients and constants (see (5.2)):

$$\delta a_{ij} = \varepsilon_{ij} \, \mathrm{sign}\!\left(\frac{\partial x_k}{\partial a_{ij}}\right) = \varepsilon_{ij} \, \mathrm{sign}(-x_j b_{ki}), \qquad i, j = 1, 2, \ldots, n,$$

$$\delta c_i = \varepsilon_i \, \mathrm{sign}\!\left(\frac{\partial x_k}{\partial c_i}\right) = \varepsilon_i \, \mathrm{sign}(b_{ki}), \qquad i = 1, 2, \ldots, n. \tag{8.2}$$

And to determine the maximum decrease in x_k we introduce changes of opposite sign to those in (8.2):

$$\delta a_{ij} = -\varepsilon_{ij} \, \mathrm{sign}\!\left(\frac{\partial x_k}{\partial a_{ij}}\right) = \varepsilon_{ij} \, \mathrm{sign}(x_j b_{ki}), \qquad i, j = 1, 2, \ldots, n,$$

$$\delta c_i = -\varepsilon_i \, \mathrm{sign}\!\left(\frac{\partial x_k}{\partial c_i}\right) = -\varepsilon_i \, \mathrm{sign}(b_{ki}), \qquad i = 1, 2, \ldots, n. \tag{8.3}$$

It is clear by (8.1), (8.2), and (8.3) that our assumption is valid as far as determining the interval of uncertainty in x_k is concerned if zero is not contained as an interior point in the intervals of uncertainty of any of the unknowns or of any of the elements in the kth row of the inverse. Then

$$U_k^{III} = U_k. \tag{8.4}$$

Otherwise, the results by Method III nevertheless correspond to changes in the unknowns that can occur within the limits of the uncertainties in the coefficients and constants so that

$$U_k^{III} \subseteq U_k \tag{8.5}$$

(see (5.78)).

Now, it will be seen that Method VI merely reduces to Method III when it can be shown by Method IV that the assumption in Method III is valid (see (5.37) and (5.38)). Thus:

$U_k^{VI} = U_k, k = 1, 2, \ldots, n$, if it can be shown by Method IV that the partial derivatives in (8.1) cannot change sign within the limits of the uncertainties in the coefficients and constants. $\qquad$ (8.6)

In particular, it will be seen that:

$U_k^{VI} = U_k$ if it can be shown by Method IV that zero is not included as an interior point in the intervals of uncertainty of any of the unknowns or of any of the elements in the kth row of the inverse. $\qquad$ (8.7)

Otherwise we shall see that

$$U_k^{VI} \supseteq U_k. \tag{8.8}$$

And it is (8.5) and (8.8) that emphasize the essential difference between Methods III and VI when the assumption in Method III is not valid.

Thus, the intervals of uncertainty by Method VI will always contain the true intervals of uncertainty and be equal to them subject to (8.7).

But how do the intervals of uncertainty by Method VI compare with those by Method IV? For by (6.18) the intervals of uncertainty by Method IV also contain the true intervals of uncertainty, i.e.,

$$U_k^{IV} = [x_k - \Delta x_k^{IV}, x_k + \Delta x_k^{IV}] \supseteq U_k = [x_k - e_k, x_k + d_k],$$
$$k = 1, 2, \ldots, n$$

(see (1.11)).

It will be seen that

$$U_k^{IV} \supseteq U_k^{VI} \supseteq U_k, \qquad k = 1, 2, \ldots, n,$$

so that the intervals by Method VI cannot be worse than those by Method IV.

We shall see further that the computational procedures for Method VI lead to two sets of intervals, namely, U_k^{VI} and U_k^{VIB} ($k = 1, 2, \ldots, n$), such that

$$U_k^{VIB} \subseteq U_k \subseteq U_k^{VI}, \qquad k = 1, 2, \ldots, n. \tag{8.9}$$

Thus, Method VI in fact gives bounds on the endpoints of the intervals of uncertainty, the endpoints being in practice sharply defined by (8.9).

8.2   Method VI

To determine the intervals of uncertainty by Method VI, we first determine the uncertainties in the unknowns and in the elements of the inverse by Method IV.

Then Method VI prescribes that in determining the interval of uncertainty in x_k we take the result by Method III if

$$\Delta x_i^{IV} \leq |x_i|, \qquad \Delta b_{ki}^{IV} \leq |b_{ki}|, \qquad i = 1, 2, \ldots, n. \tag{8.10}$$

For then the assumption in (5.1) for Method III is valid as far as determining the interval of uncertainty in x_k is concerned (see (5.37), (5.38), (5.10), and also (8.4)).

Thus

$$U_k^{VI} = U_k^{III} = U_k, \qquad \text{subject to (8.10)}. \tag{8.11}$$

But suppose not all the conditions in (8.10) are satisfied when determining the interval of uncertainty in x_k.

Then Method VI prescribes the following procedure for determining an upper bound for the maximum possible increase in x_k, i.e., for determining an upper bound for the upper endpoint of the interval of uncertainty in x_k:

1. Introduce changes δa_{ij} and δc_i determined by (8.2) and validated by (8.10), i.e., changes δa_{ij} for those i and j for which $\Delta x_j^{IV} \leq |x_j|$ and $\Delta b_{ki}^{IV} \leq |b_{ki}|$ and changes δc_i for those i for which $\Delta b_{ki}^{IV} \leq |b_{ki}|$, these changes being more specifically given in (8.12) and (8.14).

2. Put those changes equal to zero which are not validated by (8.10) but instead leave the original given uncertainties in these coefficients and constants, these changes being more specifically given in (8.13) and (8.15).

3. Solve the approximate system of equations so obtained and find the uncertainty in the kth unknown by Method IV, thereby obtaining an upper bound for the maximum increase in x_k.

Hence to obtain the maximum increase in x_k by Method VI first requires solving an approximate system of equations with the following changes δa_{ij} and δc_i prescribed by (1) and (2) above:

$$\delta a_{ij} = \varepsilon_{ij} \, \text{sign}(-x_j b_{ki}) \text{ with zero uncertainty in}$$
$$(a_{ij} + \delta a_{ij}), \text{ if both } \Delta x_j^{\text{IV}} \leq |x_j| \text{ and } \Delta b_{ki}^{\text{IV}} \leq |b_{ki}|, \quad (8.12)$$

but

$$\delta a_{ij} = 0 \text{ with an uncertainty } \varepsilon_{ij} \text{ in } a_{ij}$$
$$\text{if } \Delta x_j^{\text{IV}} > |x_j| \text{ or } \Delta b_{ki}^{\text{IV}} > |b_{ki}|, \quad (8.13)$$

and similarly

$$\delta c_i = \varepsilon_i \, \text{sign}(b_{ki}) \text{ with zero uncertainty in } (c_i + \delta c_i),$$
$$\text{if } \Delta b_{ki}^{\text{IV}} \leq |b_{ki}|, \quad (8.14)$$

but

$$\delta c_i = 0 \text{ with an uncertainty } \varepsilon_i \text{ in } c_i \text{ if } \Delta b_{ki}^{\text{IV}} > |b_{ki}|. \quad (8.15)$$

Thus to obtain the maximum increase in x_k we first require to solve the system of equations

$$\sum_{j=1}^{n} (a_{ij} + \delta a_{ij}) \bar{x}_j = c_i + \delta c_i, \qquad i = 1, 2, \ldots, n, \quad (8.16)$$

where the δa_{ij}, δc_i, and the uncertainties in the coefficients and constants are specified by (8.12) to (8.15), the solution of the system of equations in (8.16) being denoted for convenience by $\bar{x}_1, \bar{x}_2, \ldots, \bar{x}_n$.

Now, since the coefficients and constants of the system of equations in (8.16) fall within the intervals of uncertainty in the coefficients and constants of the original system of equations, it follows that $\bar{x}_k$ falls within the interval of uncertainty of x_k, i.e.,

$$\bar{x}_k \in [x_k - e_k, x_k + d_k]$$

(see (1.11)).

In particular,

$$\bar{x}_k \leq x_k + d_k, \quad (8.17)$$

i.e., $\bar{x}_k$ is a lower bound of the upper endpoint of the interval of uncertainty in x_k.

And let us emphasize that $\bar{x}_k$ is the maximum value of the kth unknown due to changes within the limits of the uncertainties in those coefficients and right-hand constants that were actually changed by (8.12) to (8.15).

Next, (3) above prescribes that the uncertainties be found by Method

IV for the system of equations in (8.16) in which the uncertainties in the coefficients and constants are given by (8.12) to (8.15), these latter uncertainties being the original uncertainties in those coefficients and constants that were not changed by (8.12) to (8.15).

If we denote the uncertainty so obtained in $\bar{x}_k$ by $\Delta\bar{x}_k^{\mathrm{IV}}$, it follows that the kth unknown cannot exceed the value

$$\bar{x}_k + \Delta\bar{x}_k^{\mathrm{IV}}.$$

For Method IV leads to upper bounds of the uncertainties (see (6.18)). It thus follows that

$$x_k + d_k \le \bar{x}_k + \Delta\bar{x}_k^{\mathrm{IV}} \tag{8.18}$$

because by (1.11) $x_k + d_k$ is the maximum value of the kth unknown for changes in the coefficients and right-hand constants within the limits of their uncertainties.

Now, combining (8.17) and (8.18) we have

$$\bar{x}_k \le x_k + d_k \le \bar{x}_k + \Delta\bar{x}_k^{\mathrm{IV}}. \tag{8.19}$$

Thus, the upper endpoint $x_k + d_k$ of the interval of uncertainty in x_k is bounded above and below in (8.19).

Summarizing, then, to obtain bounds on the maximum value of x_k by Method VI when (8.10) is not wholly satisfied, we introduce changes in some of the coefficients and right-hand constants (see (8.12) and (8.14)) but leave the original uncertainties in those coefficients and constants that are not changed (see (8.13) and (8.15)).

We now have to find the maximum decrease in x_k.

Proceeding from (8.3) (instead of (8.2) previously), and making the appropriate adjustments as regards the signs of the δa_{ij} and δc_i in (8.12) and (8.14) we obtain from (8.12) to (8.15) the approximate system of equations

$$\sum_{j=1}^{n} (a_{ij} + \delta a_{ij})\bar{\bar{x}}_j = c_i + \delta c_i, \qquad i = 1, 2, \ldots, n. \tag{8.20}$$

This leads to $\bar{\bar{x}}_k$ and $\Delta\bar{\bar{x}}_k^{\mathrm{IV}}$ such that

$$\bar{\bar{x}}_k - \Delta\bar{\bar{x}}_k^{\mathrm{IV}} \le x_k - e_k \le \bar{\bar{x}}_k \tag{8.21}$$

(see (1.11)). Thus the lower endpoint $x_k - e_k$ of the interval of uncertainty in x_k is bounded above and below by Method VI.

Now, combining the results in (8.19) and (8.21) we have for the interval of uncertainty in x_k that

$$U_k^{\mathrm{VIB}} \subseteq U_k \subseteq U_k^{\mathrm{VI}} \qquad \text{if (8.10) is not wholly satisfied} \tag{8.22}$$

where

$$U_k^{\text{VIB}} = [\bar{\bar{x}}_k, \tilde{x}_k] \tag{8.23}$$

and

$$U_k^{\text{VI}} = [\bar{\bar{x}}_k - \Delta\bar{\bar{x}}_k^{\text{IV}}, \tilde{x}_k + \Delta\tilde{x}_k^{\text{IV}}], \tag{8.24}$$

U_k^{VI} being referred to as the kth *interval of uncertainty by Method VI*.

By (8.11) and (8.22) Method VI thus leads to intervals containing the true intervals of uncertainty.

Referring to (8.2) and (8.10), let us now suppose that zero is an interior point in all the intervals of uncertainty given by Method IV for the elements in the kth row of the inverse, i.e., let us consider the case where

$$\Delta b_{ki}^{\text{IV}} > |b_{ki}|, \qquad i = 1, 2, \ldots, n. \tag{8.25}$$

Then, if we are determining the maximum change in x_k, it clearly follows by (8.12) to (8.15) that each $\delta a_{ij} = 0$ and each $\delta c_i = 0$, so we are left with the given system of equations with the given uncertainties in the coefficients and constants.

Hence, in this case

$$\tilde{x}_k = x_k = \bar{\bar{x}}_k \tag{8.26}$$

(see (8.16) and (8.20) for notation).

And since Method VI prescribes that the uncertainties be now found by Method IV we clearly have that

$$\Delta\tilde{x}_k^{\text{IV}} = \Delta x_k^{\text{IV}} = \Delta\bar{\bar{x}}_k^{\text{IV}} \tag{8.27}$$

(see (8.19) and (8.21) for notation).

Hence we have by (8.24), (8.26), and (8.27) that

$$U_k^{\text{VI}} = [x_k - \Delta x_k^{\text{IV}}, x_k + \Delta x_k^{\text{IV}}] = U_k^{\text{IV}},$$

i.e.,

$$U_k^{\text{VI}} = U_k^{\text{IV}} \qquad \text{subject to (8.25).} \tag{8.28}$$

Let us next consider the case where (8.12) to (8.15) require that actual changes be introduced in some of the coefficients and constants, the original uncertainties being left in those coefficients and constants which are not changed.

Then, since we now have a better defined problem (i.e., with fewer uncertainties in the coefficients and constants) than the one giving rise to (8.28), and since we again use Method IV to find the uncertainties, we would now expect that

$$U_k^{\text{VI}} \subseteq U_k^{\text{IV}}. \tag{8.29}$$

Now, since Δx_k^{IV} and hence U_k^{IV} is available when using Method VI we can easily compare U_k^{VI} with U_k^{IV}. Hence, in those cases where (8.29) may not be satisfied we could specify that either or both endpoints of U_k^{VI}, as necessary, be replaced by the corresponding one or ones of U_k^{IV} to ensure that (8.29) holds. Adopting this specification, then, for Method VI, (8.29) always holds.

Combining (8.22) and (8.29) we now have

$$U_k^{VIB} \subseteq U_k \subseteq U_k^{VI} \subseteq U_k^{IV}, \qquad k = 1, 2, \ldots, n. \qquad (8.30)$$

And let us recall that when (8.10) is wholly satisfied, i.e., when (8.12) to (8.15) require that all the coefficients and constants be changed to the extent of their uncertainties, then (8.11) holds, i.e.,

$$U_k^{VI} = U_k^{III} = U_k.$$

Summarizing, then, it is clear that:

The intervals of uncertainty by Method VI always contain the true intervals of uncertainty when they are not equal to them, while the intervals by Method III are only contained in the true intervals of uncertainty (8.31) when they are not equal to them (see (8.5)). At the same time, the intervals by Method VI are contained in those by Method IV, i.e., $U_k^{VI} \subseteq U_k^{IV}, k = 1, 2, \ldots, n$.

Let us now consider the volume of computation involved in applying Method VI.

In addition to the initial computation to obtain $\mathbf{x}$, $\mathbf{B}$, $\Delta \mathbf{x}^{IV}$, and $\Delta \mathbf{B}^{IV}$, Method VI requires the solution of $2n$ systems of equations to determine the $2n$ endpoints of U_k^{VI}, $k = 1, 2, \ldots, n$, the uncertainties being found by Method IV in those of the $2n$ systems where necessary. And to determine the intervals of uncertainty in the n^2 elements of the inverse Method VI would require the solution of a further $2n^2$ systems of equations.

8.3 Improved Version of Method VI

In determining the intervals of uncertainty by Method VI, we have seen that we require to know initially which intervals of uncertainty do not include zero as an interior point. For this knowledge enables one to decide by (5.32) when the assumption in (5.1) for Method III is valid in determining any particular interval of uncertainty.

Now, in Method VI we use sufficient conditions in terms of the uncertainties by Method IV to ensure that the assumption in (5.1) is valid (see (8.10) and also (5.37) and (5.38)).

But having found the intervals of uncertainty by Method VI, we have in general an improved knowledge of the intervals of uncertainty (see (8.30) and (8.31)). And making use of this improved knowledge may enable us to improve the results of Method VI.

Thus, let $\operatorname{int}(U)$ denote the interior of a closed interval U, i.e., denote the interval U with the endpoints excluded, so that while U is a closed interval, $\operatorname{int}(U)$ is the corresponding open interval. Then it might occur, for example, in view of (8.29) ($U_i^{\mathrm{VI}} \subseteq U_i^{\mathrm{IV}}$) that

$$0 \in \operatorname{int}(U_i^{\mathrm{IV}}) \qquad \text{but} \qquad 0 \notin \operatorname{int}(U_i^{\mathrm{VI}}). \tag{8.32}$$

And let us denote the interval of uncertainty in the element b_{ki} of the inverse by

$$U_{ki}, \tag{8.33}$$

the double suffix to indicate that the interval of uncertainty refers to an element of the inverse. And, further, let us denote the intervals of uncertainty in b_{ki} given by Methods IV and VI by

$$U_{ki}^{\mathrm{IV}} \qquad \text{and} \qquad U_{ki}^{\mathrm{VI}} \tag{8.34}$$

respectively.

Then, corresponding to (8.30) we have for the intervals of uncertainty of the elements of the inverse that

$$U_{ki}^{\mathrm{VIB}} \subseteq U_{ki} \subseteq U_{ki}^{\mathrm{VI}} \subseteq U_{ki}^{\mathrm{IV}}, \qquad i, k = 1, 2, \ldots, n, \tag{8.35}$$

where the interval U_{ki}^{VIB} for the element b_{ki} corresponds to the interval U_k^{VIB} in (8.23) for the unknown x_k.

Hence, it is possible that

$$0 \in \operatorname{int}(U_{ki}^{\mathrm{IV}}) \qquad \text{but} \qquad 0 \notin \operatorname{int}(U_{ki}^{\mathrm{VI}}). \tag{8.36}$$

Now, it is clear by (5.32) that the conditions

$$0 \notin \operatorname{int}(U_i), \qquad 0 \notin \operatorname{int}(U_{ki}), \qquad i = 1, 2, \ldots, n, \tag{8.37}$$

will ensure that the interval of uncertainty in x_k given by Method III is the true interval of uncertainty.

And the sufficient conditions in (8.10) to ensure that (8.37) holds may be put in the form

$$0 \notin \operatorname{int}(U_i^{\mathrm{IV}}), \qquad 0 \notin \operatorname{int}(U_{ki}^{\mathrm{IV}}), \qquad i = 1, 2, \ldots, n. \tag{8.38}$$

But in view of (8.30) and (8.35) it is clear that a less demanding set of sufficient conditions to ensure that (8.37) holds, i.e., to ensure that the interval of uncertainty in x_k given by Method III is the true interval of uncertainty, is

$$0 \notin \text{int}(U_i^{\text{VI}}), \qquad 0 \notin \text{int}(U_{ki}^{\text{VI}}), \qquad i = 1, 2, \ldots, n, \qquad (8.39)$$

(see also (8.32) and (8.36)).

Let us then rewrite the specifications for Method VI in (8.12) to (8.15) so as to include the less stringent conditions in (8.39) as well as those in (8.38), i.e., those in (8.10). Thus, (8.12) to (8.15) become:

> Put $\delta a_{ij} = \varepsilon_{ij} \, \text{sign}(-x_j b_{ki})$ with zero uncertainty in $(a_{ij} + \delta a_{ij})$ if both (a) and (b) hold:
>
> (a) $0 \notin \text{int}(U_j^{\text{IV}})$ or $0 \notin \text{int}(U_j^{\text{VI}})$
>
> (b) $0 \notin \text{int}(U_{ki}^{\text{IV}})$ or $0 \notin \text{int}(U_{ki}^{\text{VI}})$,   (8.40)

but

> put $\delta a_{ij} = 0$ with an uncertainty ε_{ij} in a_{ij} if either (a) or (b) in (8.40) does not hold,   (8.41)

and similarly

> put $\delta c_i = \varepsilon_i \, \text{sign}(b_{ki})$ with zero uncertainty in $(c_i + \delta c_i)$ if $0 \notin \text{int}(U_{ki}^{\text{IV}})$ or $0 \notin \text{int}(U_{ki}^{\text{VI}})$,   (8.42)

but

> put $\delta c_i = 0$ with an uncertainty ε_i in c_i if neither condition in (8.42) holds.   (8.43)

Then, the improved version of Method VI involves replacing the original specifications in (8.12) to (8.15) for Method VI by those in (8.40) to (8.43). The new specifications imply that we use the new and in general improved information regarding the intervals of uncertainty as it becomes available in place of the initial information given by Method IV. And in the improved version we assume that we continue rerunning the whole or relevant part of the procedure until no further improvement is obtainable.

This then constitutes the improved version of Method VI.

Let us now note that if zero is an internal point of U_j^{VIB} then no further improvement is possible in this respect in the specifications (8.40) to (8.43) on rerunning the procedure because by (8.22)

$$U_j^{\text{VIB}} \subseteq U_j \subseteq U_j^{\text{VI}}.$$

And similar considerations hold if zero is an internal point of U_{ki}^{VIB}, say.

Let us note, further, that in using the improved version of Method VI for finding the uncertainties in the unknowns one could restrict oneself to using only new information regarding the intervals of uncertainty in the unknowns, no new information being obtained regarding the intervals of uncertainty in the elements of the inverse. For the volume of computation to find the U_{ki}^{VI} ($k, i = 1, 2, \ldots, n$) can be large. But of course the results may not be as good as otherwise obtainable.

Finally, let us note that it may be expected that the results of the ordinary version of Method VI or those obtained in the first run of the improved version of Method VI would frequently not be able to be improved further or in any case would be sufficiently good. This is especially so since the improvements are likely to be small. For the partial derivative

$$\frac{\partial x_k^*}{\partial a_{ij}^*} = -x_j^* b_{ki}^*, \tag{8.44}$$

for example, may well be small if zero is included in the interval of uncertainty of x_j or b_{ki} by Method IV.

8.4 Example

For our example in (1.43) and (1.44), we have by (5.64), (6.71), (5.67), and (5.69) that

$$x_1 = 0\cdot9730, \qquad x_2 = -0\cdot9962, \qquad x_3 = 2\cdot0438 \tag{8.45}$$

$$\Delta x_1^{IV} = 0\cdot1561, \qquad \Delta x_2^{IV} = 0\cdot0763, \qquad \Delta x_3^{IV} = 0\cdot2400 \tag{8.46}$$

$$\mathbf{B} = \begin{bmatrix} -1\cdot0574 & 0\cdot0724 & 1\cdot7153 \\ 0\cdot8878 & -0\cdot5017 & 0\cdot0013 \\ 1\cdot6603 & 0\cdot8189 & -1\cdot8963 \end{bmatrix} \tag{8.47}$$

$$\Delta\mathbf{B}^{IV} = \begin{bmatrix} 0\cdot1122 & 0\cdot0434 & 0\cdot1125 \\ 0\cdot0549 & 0\cdot0212 & 0\cdot0550 \\ 0\cdot1726 & 0\cdot0667 & 0\cdot1730 \end{bmatrix}. \tag{8.48}$$

Thus, the conditions in (8.10) are all satisfied except that

$$\Delta b_{23}^{IV} = 0\cdot0550 \nleqslant 0\cdot0013 = |b_{23}|. \tag{8.49}$$

This would imply by (8.11) that Method VI can only differ from Method III with respect to the interval of uncertainty in x_2, since by (8.11)

$$U_1^{VI} = U_1^{III} = U_1 \quad \text{and} \quad U_3^{VI} = U_3^{III} = U_3. \tag{8.50}$$

Now, by (1.43), (1.44), (8.45) to (8.48), and (8.12) to (8.15) the system of equations to be solved to determine the maximum value of x_2 is

$$(0.53-0.01)\bar{x}_1+(0.86+0.01)\bar{x}_2+(0.48-0.01)\bar{x}_3 = 0.64+0.01$$
$$(0.94+0.01)\bar{x}_1+(-0.47-0.01)\bar{x}_2+(0.85+0.01)\bar{x}_3 = 3.12-0.01 \quad (8.51)$$
$$(0.87+0)\bar{x}_1+(0.55+0)\bar{x}_2+(0.26+0)\bar{x}_3 = 0.83+0$$

or, in matrix notation,

$$\bar{\mathbf{A}}\bar{\mathbf{x}} = \bar{\mathbf{c}}. \qquad (8.52)$$

And by (8.12) to (8.15) the uncertainties in the coefficient matrix and in the column of constants in (8.51) are

$$\Delta\bar{\mathbf{A}} = \begin{bmatrix} 0 & 0 & 0 \\ 0 & 0 & 0 \\ 0.01 & 0.01 & 0.01 \end{bmatrix}, \quad \Delta\bar{\mathbf{c}} = \begin{bmatrix} 0 \\ 0 \\ 0.01 \end{bmatrix}. \qquad (8.53)$$

Now, the solution of (8.51) is

$$\bar{x}_1 = 0.91729258, \quad \bar{x}_2 = -0.92764644, \quad \bar{x}_3 = 2.08523459 \qquad (8.54)$$

and the inverse of the coefficient matrix is

$$\bar{\mathbf{B}} = \begin{bmatrix} -1.0542 & 0.0569 & 1.7173 \\ 0.8838 & -0.4826 & -0.0012 \\ 1.6579 & 0.8304 & -1.8977 \end{bmatrix}, \qquad (8.55)$$

the values given being obtained by truncating more accurate ones.

Hence, the ill-conditioning matrix

$$\bar{\mathbf{C}} = |\bar{\mathbf{B}}|\,\Delta\bar{\mathbf{A}} = \begin{bmatrix} 1.0542 & 0.0569 & 1.7173 \\ 0.8838 & 0.4826 & 0.0012 \\ 1.6579 & 0.8304 & 1.8977 \end{bmatrix}\begin{bmatrix} 0 & 0 & 0 \\ 0 & 0 & 0 \\ 0.01 & 0.01 & 0.01 \end{bmatrix}$$

$$= \begin{bmatrix} 0.0171 & 0.0171 & 0.0171 \\ 0.0000 & 0.0000 & 0.0000 \\ 0.0189 & 0.0189 & 0.0189 \end{bmatrix}. \qquad (8.56)$$

It follows by (2.24) that

$$\rho(\bar{\mathbf{C}}) \le \|\bar{\mathbf{C}}\|_{\mathrm{II}} = 0.0171 + 0.0000 + 0.0189 < 1.$$

Hence, we have by (6.18), (6.36), (4.23), and (8.53) to (8.56) that

$$\Delta\bar{\mathbf{x}}^{\mathrm{IV}} = (\mathbf{I}-\bar{\mathbf{C}})^{-1}\,\Delta\bar{\mathbf{x}}^{\mathrm{II}} = (\mathbf{I}-\bar{\mathbf{C}})^{-1}\,(\bar{\mathbf{C}}\,|\bar{\mathbf{x}}| + |\bar{\mathbf{B}}|\,\Delta\bar{\mathbf{c}})$$

$$= \begin{bmatrix} 0.08784534 \\ 0.00006314 \\ 0.09707371 \end{bmatrix} \ge \Delta\bar{\mathbf{x}}. \qquad (8.57)$$

Hence, by (8.54), (8.57), and (8.19) we have for $k = 2$ that

$$\bar{x}_2 = -0{\cdot}92764644 \leq x_2 + d_2 \leq \bar{x}_2 + \Delta\bar{x}_2^{IV} = -0{\cdot}92764644 + 0{\cdot}00006314$$

i.e.,

$$-0{\cdot}92764644 \leq x_2 + d_2 \leq -0{\cdot}92758329. \qquad (8.58)$$

This, then, bounds the upper endpoint $x_2 + d_2$ of the interval of uncertainty in x_2.

And to obtain a lower bound for the lower endpoint of the interval of uncertainty in x_2 we have to introduce changes in the coefficients and constants of opposite sign to those introduced in (8.51). This leads to

$$(0{\cdot}53+0{\cdot}01)\bar{\bar{x}}_1+(0{\cdot}86-0{\cdot}01)\bar{\bar{x}}_2+(0{\cdot}48+0{\cdot}01)\bar{\bar{x}}_3 = 0{\cdot}64-0{\cdot}01$$
$$(0{\cdot}94-0{\cdot}01)\bar{\bar{x}}_1+(-0{\cdot}47+0{\cdot}01)\bar{\bar{x}}_2+(0{\cdot}85-0{\cdot}01)\bar{\bar{x}}_3 = 3{\cdot}12+0{\cdot}01 \quad (8.59)$$
$$(0{\cdot}87+0)\bar{\bar{x}}_1+(0{\cdot}55+0)\bar{\bar{x}}_2+(0{\cdot}26+0)\bar{\bar{x}}_3 = 0{\cdot}83+0$$

or, in matrix notation,

$$\bar{\bar{A}}\bar{\bar{x}} = \bar{\bar{c}}. \qquad (8.60)$$

The uncertainties in the coefficients and constants in (8.59) are, however, the same as those in (8.53).

Proceeding as before, we now obtain

$$\bar{\bar{x}}_1 = 1{\cdot}03060262, \quad \bar{\bar{x}}_2 = -1{\cdot}06699567, \quad \bar{\bar{x}}_3 = 2{\cdot}00085898 \quad (8.61)$$

and

$$\Delta\bar{\bar{x}}^{IV} = (I - \bar{\bar{C}})^{-1}\,\Delta\bar{\bar{x}}^{II} = \begin{bmatrix} 0{\cdot}09062166 \\ 0{\cdot}00020258 \\ 0{\cdot}10022018 \end{bmatrix} \geq \Delta\bar{\bar{x}}. \qquad (8.62)$$

Hence, by (8.21) we have

$$\bar{\bar{x}}_2 - \Delta\bar{\bar{x}}_2^{IV} = -1{\cdot}06719825 \leq x_2 - e_2 \leq \bar{\bar{x}}_2 = -1{\cdot}06699567$$

i.e.

$$-1{\cdot}06719825 \leq x_2 - e_2 \leq -1{\cdot}06699567. \qquad (8.63)$$

We have thus bounded the lower endpoint $x_2 - e_2$ of the interval of uncertainty in x_2.

Combining the results in (8.58) and (8.63) we then have that

$$U_2^{VIB} \subseteq U_2 \subseteq U_2^{VI}$$

where

$$U_2^{VIB} = [-1{\cdot}06699567, \quad -0{\cdot}92764644]$$
$$U_2^{VI} = [-1{\cdot}06719825, \quad -0{\cdot}92758329] \qquad (8.64)$$

(see (8.30), (8.23), and (8.24)).

Clearly, we may regard U_2 as sharply defined by U_2^{VI} and U_2^{VIB}, the difference in widths between the intervals being

$$w(U_2^{\text{VI}}) - w(U_2^{\text{VIB}}) = 0{\cdot}13961495 - 0{\cdot}13934923 = 0{\cdot}00026572. \quad (8.65)$$

(Note that this value is in fact equal to

$$\Delta \bar{x}_2^{\text{IV}} + \Delta \bar{\bar{x}}_2^{\text{IV}} = 0{\cdot}00006314 + 0{\cdot}00020258 = 0{\cdot}00026572$$

(see (8.23) and (8.24)).)

Let us now compare the values in (8.64) with those obtainable by combining Methods IV and V.

By (6.18) and (7.7) we have

$$U_2^{\text{V}} = [x_2 - \Delta x_2^{\text{V}}, \, x_2 + \Delta x_2^{\text{V}}] \subseteq U_2 \subseteq [x_2 - \Delta x_2^{\text{IV}}, \, x_2 + \Delta x_2^{\text{IV}}]. \quad (8.66)$$

Hence, with the values in (8.45), (8.46), and (7.51) we have

$$U_2^{\text{V}} = [-1{\cdot}0593, \, -0{\cdot}9331] \subseteq U_2 \subseteq [-1{\cdot}0725, \, 0{\cdot}9199] = U_2^{\text{IV}}, \quad (8.67)$$

the difference in widths between the intervals U_2^{IV} and U_2^{V} being

$$w(U_2^{\text{IV}}) - w(U_2^{\text{V}}) = 0{\cdot}1526 - 0{\cdot}1262 = 0{\cdot}0264. \quad (8.68)$$

Thus, Method VI defines the interval of uncertainty in x_2 more sharply than Methods IV and V (see (8.65)). And this improvement holds both for the intervals containing and contained by the true interval of uncertainty. For, clearly, by (8.64) and (8.67)

$$U_2^{\text{V}} \subset U_2^{\text{VIB}} \subseteq U_2 \subseteq U_2^{\text{VI}} \subset U_2^{\text{IV}}. \quad (8.69)$$

In conclusion, we may say that one may regard Method VI as the practical application of Method III. For Method VI reduces to Method III when the relevant portion of the assumption of Method III in (5.1) can be shown to be valid. And when it cannot be shown to be valid we obtain intervals of uncertainty containing the true intervals rather than being contained in them.

Loosely speaking, one may say that the results by Method VI always fall between the true results and those by Method IV. Thus, Method VI is the obvious procedure when one wants to go one step better than Method IV.

We may mention that we shall deal with an interval arithmetic version of Method VI in the next chapter in Section 9.8.

CHAPTER 9

Methods VII and VIII: Uncertainties by Interval Arithmetic

9.1 Introduction

Interval arithmetic can be used in general in any arithmetical procedure in which there are uncertainties in the given values. And if *rounded interval arithmetic* (discussed in Section 9.2) is used during the computation then the uncertainties in the final results will always be upper bounds.

It will thus become clear that:

> Rounded interval arithmetic leads to upper bounds of the uncertainties. (9.1)

Now, it will be seen in Section 9.4 that the results by interval arithmetic depend on the precise arithmetical steps. Two procedures equivalent mathematically may lead to different final intervals. Thus, while we find that the intervals of uncertainty obtained in Section 9.3 by solving our numerical example to be more than twice as large as those by Method IV, it will be seen that we obtain for the same example intervals of uncertainty of smaller width than those by Method IV if we carry out the transformation suggested by Hansen (reference 2) before proceeding to the solution (Section 9.5). We call this latter procedure our Method VII.

Now, in Sections 9.3 and 9.5 we only consider the numerical determination of the uncertainties by interval arithmetic, these uncertainties being upper bounds by (9.1).

But interval arithmetic can also be used to obtain theoretical upper bounds for the uncertainties. These upper bounds of the uncertainties which we obtain in Section 9.7 are in fact those by Method IV.

And in Section 9.8 we deal with the interval arithmetic version of Method VI, Method IV used in Method VI being replaced by Method VII. We call this our Method VIII (see reference 3).

9.2 Interval Arithmetic

Definitions

An *interval number* can be considered either as an ordered pair of real numbers $[a, b]$, with $a \leq b$, or preferably as a set of real numbers. Thus $[a, b]$ can be considered as the set of real numbers x such that $a \leq x \leq b$, or briefly,

$$[a, b] = \{x \mid a \leq x \leq b\} \qquad (9.2)$$

The following results, for example, then follow from set theory:

$$\begin{aligned}
[a, b] &= [c, d] && \text{iff (if and only if)} \quad a = c, b = d \\
[a, b] &\subseteq [c, d] && \text{iff} \quad c \leq a \leq b \leq d \\
[a, b] &< [c, d] && \text{iff} \quad b < c.
\end{aligned} \qquad (9.3)$$

And we denote the *width* of an interval $[a, b]$ by

$$w([a, b]) = b - a \qquad (9.4)$$

and the *magnitude* of the interval by

$$|[a, b]| = \max(|a|, |b|). \qquad (9.5)$$

Let us note that *degenerate intervals*, i.e., intervals of zero width, can be identified with real numbers. Thus $a = [a, a]$.

We now define the arithmetic operations on intervals, it being assumed that the endpoints are computed with infinite precision.

If * is one of the symbols $+$, $-$, $.$, $/$, then the interval

$$[a, b] * [c, d] = \{x * y \mid a \leq x \leq b, c \leq y \leq d\} \qquad (9.6)$$

except that we do not define $[a, b]/[c, d]$ if $0 \in [c, d]$. (This excludes the possibility of division by zero.)

Alternatively, we can define the arithmetic operations on intervals by giving the endpoints of the intervals resulting from the sum, difference, product, or quotient of two intervals. Equivalent to the definition in (9.6) we thus have

$$\begin{aligned}
[a, b] + [c, d] &= [a + c, b + d] \\
[a, b] - [c, d] &= [a - d, b - c] \\
[a, b] \ . \ [c, d] &= [\min(ac, ad, bc, bd), \max(ac, ad, bc, bd)] \\
[a, b] \ / \ [c, d] &= [a, b] \ . \ [1/d, 1/c], \qquad \text{provided } 0 \notin [c, d],
\end{aligned} \qquad (9.7)$$

for the denominator cannot be zero. Thus, in the definition of division c and d are either both positive or both negative.

By way of a numerical example we thus have

$$[-2, 3] + [-5, -4] = [-2 - 5, 3 - 4] = [-7, -1]$$
$$[-2, 3] - [-5, -4] = [-2 - (-4), 3 - (-5)] = [2, 8]$$
$$[-2, 3] \cdot [-5, -4] = [\min(10, 8, -15, -12),$$
$$\max(10, 8, -15, -12)]$$
$$= [-15, 10]$$
$$[-2, 3] / [-5, -4] = [\min(1/2, 2/5, -3/4, -3/5),$$
$$\max(1/2, 2/5, -3/4, -3/5)]$$
$$= [-3/4, 1/2].$$

(9.8)

Properties of Interval Arithmetic

We may note that interval arithmetic is a generalization or extension of real arithmetic, since $[a, a]$ is a real number and the definitions hold for intervals of this form as well.

Thus, laws that hold for interval arithmetic must automatically hold for real arithmetic. But laws that hold for real arithmetic may not always hold for interval arithmetic.

For example, interval arithmetic is *associative* and *commutative* with respect to addition and multiplication. Thus, given interval numbers I, J, K we have

$$I + (J + K) = (I + J) + K$$
$$I + J = J + I.$$

But the *distributive* law does not always hold for interval arithmetic. Thus, while

$$[1, 2]([3, 4] + [-6, -5]) = [1, 2] \cdot [-3, -1] = [-6, -1], \quad (9.9)$$

distributively it would be

$$[1, 2] \cdot [3, 4] + [1, 2] \cdot [-6, -5]$$
$$= [3, 8] + [-12, -5] = [-9, 3]. \quad (9.10)$$

But the distributive law does sometimes hold. For example, if k is a real number, i.e., the degenerate interval $[k, k]$ of zero width, then it is easy to show that

$$k([a, b] + [c, d]) = k[a, b] + k[c, d] \quad (9.11)$$

holds for all real k.

It thus follows, for example, that

$$b[a - \delta a, a + \delta a] = b([a, a] + [-\delta a, \delta a])$$
$$= b[a, a] + b[-\delta a, \delta a]$$
$$= [ba, ba] + [-|b| \, \delta a, |b| \, \delta a]$$
$$= [ba - |b| \, \delta a, ba + |b| \, \delta a],$$

i.e.,

$$b[a - \delta a, a + \delta a] = [ba - |b| \, \delta a, ba + |b| \, \delta a]. \qquad (9.12)$$

(We may note that δa is nonnegative by implication of its appearance in $[a - \delta a, a + \delta a]$, for in any interval $[c, d]$, $c \leq d$.)

It is not our intention to discuss the properties of interval arithmetic extensively here. For a fuller discussion we refer the reader to reference 4.

Rounded Interval Arithmetic

An important use of interval arithmetic is to determine upper bounds for the errors due to rounding.

In *rounded interval arithmetic* working in either single or double precision, we regard each real number as an interval. And in the first place we round the lower endpoint down and the upper endpoint up.

All arithmetic operations are then carried out on intervals, at each stage the lower endpoint being rounded down and the upper endpoint being rounded up.

In this way, the maximum possible effect of rounding is taken into account, so that the interval for the final result contains the true value.

In using rounded interval arithmetic it should be clear that the final result can be obtained as sharply as desired, i.e., the width of the final interval can be as small as desired, by using a sufficiently long word-length.

By way of a numerical example suppose we multiply the two numbers

$$a = 0 \cdot 3678 \mid 246 \ldots \qquad \text{and} \qquad b = 0 \cdot 9367 \mid 847 \ldots \qquad (9.13)$$

in a decimal coded machine using a precision or wordlength of four decimal digits.

Then, in the first place we write down the intervals containing a and b:

$$a \in [0 \cdot 3678, 0 \cdot 3679], \qquad b \in [0 \cdot 9367, 0 \cdot 9368]. \qquad (9.14)$$

Then, carrying out the multiplication we have

$$ab \in [0 \cdot 3678, 0 \cdot 3679] \, . \, [0 \cdot 9367, 0 \cdot 9368]$$
$$= [0 \cdot 3445 | 18 \ldots, 0 \cdot 3446 | 48 \ldots] \subseteq [0 \cdot 3445, 0 \cdot 3447].$$

Thus, rounded interval arithmetic leads to the following interval which contains ab:

$$[0 \cdot 3445, 0 \cdot 3447], \qquad (9.15)$$

the width of this interval being $0 \cdot 3447 - 0 \cdot 3445 = 0 \cdot 0002$.

To obtain the product ab more sharply, we clearly have to use a longer wordlength for the computation.

The actual rounding down and up can most easily be achieved in machine coding. The lower endpoint should be rounded down so that the rounded value is less than or equal to the value being rounded. Thus, if the value being rounded down is positive we need merely truncate while if the value is negative we truncate, examine the truncated portion, and, if this is not zero, we subtract one unit in the least significant position. And the upper endpoint must be rounded up if a computation is being carried out in rounded interval arithmetic, an interval being said to be *properly rounded* if the endpoints are rounded as described above.

We may mention that the rounding up and down can be coded in Fortran if necessary, although not as efficiently.

Summarizing, then, the final interval obtained by rounded interval arithmetic takes account of the errors due to rounding and contains the true value.

Interval Vectors and Matrices

Interval vectors and matrices are vectors and matrices whose elements are interval numbers, the superscript I being used to indicate such a vector or matrix.

Thus given matrices $\mathbf{B} = (b_{ij})$ and $\mathbf{C} = (c_{ij})$ of order n such that

$$b_{ij} \leq c_{ij}, \qquad i, j = 1, 2, \ldots, n, \tag{9.16}$$

then the interval matrix $\mathbf{A}^I = [\mathbf{B}, \mathbf{C}]$ is defined by

$$\mathbf{A}^I = [\mathbf{B}, \mathbf{C}] = \{\mathbf{A} = (a_{ij}) \mid b_{ij} \leq a_{ij} \leq c_{ij}, \quad i, j = 1, 2, \ldots, n\}. \tag{9.17}$$

And let us conveniently call $\mathbf{B}$ the *lower endpoint matrix* of the interval matrix $\mathbf{A}^I = [\mathbf{B}, \mathbf{C}]$, $\mathbf{C}$ the *upper endpoint matrix*, and $\mathbf{M} = \frac{1}{2}(\mathbf{B} + \mathbf{C})$ the *midpoint matrix*.

Suppose we write

$$\mathbf{B} \leq \mathbf{C} \qquad \text{iff} \quad b_{ij} \leq c_{ij}, \qquad i, j = 1, 2, \ldots, n,$$

and $\tag{9.18}$

$$\mathbf{B} \geq \mathbf{C} \qquad \text{iff} \quad b_{ij} \geq c_{ij}, \qquad i, j = 1, 2, \ldots, n.$$

Then our definition for an interval matrix can be briefly written as

$$\mathbf{A}^I = [\mathbf{B}, \mathbf{C}] = \{\mathbf{A} \mid \mathbf{B} \leq \mathbf{A} \leq \mathbf{C}\}. \tag{9.19}$$

By way of notation suppose now that

$$\mathbf{A}^I = [\mathbf{B}, \mathbf{C}], \qquad \mathbf{D}^I = [\mathbf{E}, \mathbf{F}]. \tag{9.20}$$

Then we have that

$$\mathbf{A} \in \mathbf{A}^I = [\mathbf{B}, \mathbf{C}] \qquad \text{iff} \quad \mathbf{B} \leq \mathbf{A} \leq \mathbf{C}, \tag{9.21}$$

$$\mathbf{A}^I \subseteq \mathbf{D}^I \qquad \text{iff} \quad \mathbf{E} \leq \mathbf{B} \leq \mathbf{C} \leq \mathbf{F}, \tag{9.22}$$

and

$$\mathbf{A}^I = \mathbf{D}^I \qquad \text{iff} \quad \mathbf{B} = \mathbf{E}, \quad \text{and} \quad \mathbf{C} = \mathbf{F}. \tag{9.23}$$

And as with interval numbers a real matrix $\mathbf{A}$ can be regarded as a degenerate interval matrix $[\mathbf{A}, \mathbf{A}]$.

Let us now define operations on square interval matrices:

$$\mathbf{A}^I + \mathbf{B}^I = \{\mathbf{A} + \mathbf{B} \mid \mathbf{A} \in \mathbf{A}^I, \mathbf{B} \in \mathbf{B}^I\} \tag{9.24}$$

$$\mathbf{A}^I - \mathbf{B}^I = \{\mathbf{A} - \mathbf{B} \mid \mathbf{A} \in \mathbf{A}^I, \mathbf{B} \in \mathbf{B}^I\} \tag{9.25}$$

$$\mathbf{A}^I\mathbf{B}^I = \{\mathbf{AB} \mid \mathbf{A} \in \mathbf{A}^I, \mathbf{B} \in \mathbf{B}^I\} \tag{9.26}$$

$$(\mathbf{A}^I)^{-1} = \{\mathbf{A}^{-1} \mid \mathbf{A} \in \mathbf{A}^I\} \tag{9.27}$$

except that we do not define $(\mathbf{A}^I)^{-1}$ if $\det(\mathbf{A}) = 0$ for any $\mathbf{A} \in \mathbf{A}^I$.

In the case of interval matrices of order m, n where $m \neq n$, and this includes vectors, results corresponding to those from (9.16) to (9.26) hold, but the inverse concept defined in (9.27) of course does not hold.

By way of example, if

$$[-\Delta\mathbf{A}, \Delta\mathbf{A}] \tag{9.28}$$

is an interval matrix then $\Delta\mathbf{A}$ is clearly a nonnegative matrix by virtue of its appearance in the interval.

Clearly, then, for any matrix $\mathbf{B}$ of suitable order

$$\mathbf{B}[-\Delta\mathbf{A}, \Delta\mathbf{A}] \subseteq [-|\mathbf{B}| \Delta\mathbf{A}, |\mathbf{B}| \Delta\mathbf{A}] \tag{9.29}$$

and

$$[-\Delta\mathbf{A}, \Delta\mathbf{A}]\mathbf{B} \subseteq [-\Delta\mathbf{A} |\mathbf{B}|, \Delta\mathbf{A} |\mathbf{B}|]. \tag{9.30}$$

And, corresponding to (9.12), it is easy to show that

$$\mathbf{B}[\mathbf{A} - \Delta\mathbf{A}, \mathbf{A} + \Delta\mathbf{A}] \subseteq [\mathbf{BA} - |\mathbf{B}| \Delta\mathbf{A}, \mathbf{BA} + |\mathbf{B}| \Delta\mathbf{A}]. \tag{9.31}$$

And similarly it can be shown that

$$[\mathbf{A} - \Delta\mathbf{A}, \mathbf{A} + \Delta\mathbf{A}]\mathbf{B} \subseteq [\mathbf{AB} - \Delta\mathbf{A} |\mathbf{B}|, \mathbf{AB} + \Delta\mathbf{A} |\mathbf{B}|]. \tag{9.32}$$

Having dealt briefly with some theoretical aspects of interval arithmetic, we now proceed to apply interval arithmetic techniques to the solution of approximate linear algebraic equations.

9.3 Ordinary Method Interval Arithmetic

We shall show in Section 9.4 that the results obtained using interval arithmetic depend on the precise arithmetical steps and that two procedures equivalent mathematically may lead to different final intervals.

It may, therefore, be necessary to choose arithmetical procedures which would tend to lead to small intervals in the final results.

But first we merely consider our normal solution procedure, namely Gaussian elimination using partial pivoting, and apply rounded interval arithmetic during the computation. This leads to the heading of this section: ordinary method interval arithmetic. And we refer to the results obtained in this way as the results by O.M.

Corresponding to the form in Table 1.2, we give the results presently for our example in (1.43) and (1.44), rounded interval arithmetic being used throughout the solution procedure and the inverse being found at the same time as in Table 1.3.

The computation was carried out in a wordlength of eight decimal digits and at each stage the endpoints were *properly rounded*, so that at all stages the computed intervals contain the true intervals of uncertainty without doubt.

The results given below are, however, our 8S (eight significant figure) results truncated to 4D (four decimal places) and the values must be adjusted to obtain properly rounded results.

The solution of (1.43) for an uncertainty of 0·01 in each coefficient and constant, using rounded interval arithmetic in the Gaussian elimination procedure with partial pivoting, this leading to the so-called O.M. (ordinary method) results.

0·5300	0·8600	0·4800	0·6400	1·0000	0·0000	0·0000
0·9400	−0·4700	0·8500	3·1200	0·0000	1·0000	0·0000
0·8700	0·5500	0·2600	0·8300	0·0000	0·0000	1·0000

 Coefficient matrix   Constants   Unit matrix

The given system of equations; refer to the first three rows in Table 1.3.

0·01	0·01	0·01	0·01
0·01	0·01	0·01	0·01
0·01	0·01	0·01	0·01

The matrix of uncertainties in the coefficients and right-hand constants.

Step 1

0·5400	0·8700	0·4900	0·6500	1·0000	0·0000	0·0000
0·9500	−0·4600	0·8600	3·1300	0·0000	1·0000	0·0000
0·8800	0·5600	0·2700	0·8400	0·0000	0·0000	1·0000

0·5200	0·8500	0·4700	0·6300	1·0000	0·0000	0·0000
0·9300	−0·4800	0·8400	3·1100	0·0000	1·0000	0·0000
0·8600	0·5400	0·2500	0·8200	0·0000	0·0000	1·0000

The endpoints of the intervals of the given system of equations. We adopt the procedure of giving the upper endpoints in the first three rows and the lower endpoints in the last three rows. Thus, the first three rows constitute the *upper endpoint matrix* and the last three rows the *lower endpoint matrix*. Note that the unit matrices in the last three columns are not affected at this stage because their elements are known exactly.

Step 2

0·9500	−0·4600	0·8600	3·1300	0·0000	1·0000	0·0000
0·5400	0·8700	0·4900	0·6500	1·0000	0·0000	0·0000
0·8800	0·5600	0·2700	0·8400	0·0000	0·0000	1·0000
0·9300	−0·4800	0·8400	3·1100	0·0000	1·0000	0·0000
0·5200	0·8500	0·4700	0·6300	1·0000	0·0000	0·0000
0·8600	0·5400	0·2500	0·8200	0·0000	0·0000	1·0000

The first and second rows are interchanged in both the upper and lower endpoint matrices, so that the element of largest magnitude in each first column is in the first row. It should be noted that in deciding which rows to interchange for our partial pivoting procedure we referred only to the elements in the upper endpoint matrix. (Note that with the figures in Table 1.2 no interchange of rows was necessary.)

Step 3

1·0000	−0·4842	0·9247	3·3655	0·0000	1·0752	0·0000
0·5400	0·8700	0·4900	0·6500	1·0000	0·0000	0·0000
0·8800	0·5600	0·2700	0·8400	0·0000	0·0000	1·0000
1·0000	−0·5161	0·8842	3·2736	0·0000	1·0526	0·0000
0·5200	0·8500	0·4700	0·6300	1·0000	0·0000	0·0000
0·8600	0·5400	0·2500	0·8200	0·0000	0·0000	1·0000

To produce 1·0000 in the pivotal position, the first pivotal row is divided by the pivot, which is the interval [0·93, 0·95], rounded interval arithmetic being used. Note, however, that 1·0000 is placed in the pivotal position of each endpoint matrix because the purpose of the division was to achieve this effect.

Step 4

1·0000	−0·4842	0·9247	3·3655	0·0000	1·0752	0·0000
0·0000	1·1487	0·0302	−1·0523	1·0000	−0·5473	0·0000
0·8800	0·5600	0·2700	0·8400	0·0000	0·0000	1·0000
1·0000	−0·5161	0·8842	3·2736	0·0000	1·0526	0·0000
0·0000	1·1017	−0·0293	−1·1874	1·0000	−0·5806	0·0000
0·8600	0·5400	0·2500	0·8200	0·0000	0·0000	1·0000

Rounded interval arithmetic is used and a multiple of the first pivotal row is subtracted from the row below it to produce a zero below the pivot. Note that we place a zero in this position below the pivot because this was the required effect.

Step 5

1·0000	−0·4842	0·9247	3·3655	0·0000	1·0752	0·0000
0·0000	1·1487	0·0302	−1·0523	1·0000	−0·5473	0·0000
0·0000	1·0141	−0·4904	−1·9753	0·0000	−0·9052	1·0000
1·0000	−0·5161	0·8842	3·2736	0·0000	1·0526	0·0000
0·0000	1·1017	−0·0293	−1·1874	1·0000	−0·5806	0·0000
0·0000	0·9564	−0·5637	−2·1417	0·0000	−0·9462	1·0000

A multiple of the pivotal row is subtracted from the third row to produce a zero in the first column of this row, rounded interval arithmetic being used. As was indicated earlier, we give for convenience the upper endpoints of the interval matrix in the first three rows and the lower endpoints in the last three rows.

Step 6

1·0000	−0·4842	0·9247	3·3655	0·0000	1·0752	0·0000
0·0000	1·0000	0·0274	−0·9160	0·9076	−0·4765	0·0000
0·0000	1·0141	−0·4904	−1·9753	0·0000	−0·9052	1·0000
1·0000	−0·5161	0·8842	3·2736	0·0000	1·0526	0·0000
0·0000	1·0000	−0·0266	−1·0777	0·8705	−0·5270	0·0000
0·0000	0·9564	−0·5637	−2·1417	0·0000	−0·9462	1·0000

The second pivotal row of the interval matrix is divided by the pivot in rounded interval arithmetic, 1·0000 being placed in the pivotal position.

Step 7

1·0000	−0·4842	0·9247	3·3655	0·0000	1·0752	0·0000
0·0000	1·0000	0·0274	−0·9160	0·9076	−0·4765	0·0000
0·0000	0·0000	−0·4634	−0·8823	−0·8326	−0·3707	1·0000
1·0000	−0·5161	0·8842	3·2736	0·0000	1·0526	0·0000
0·0000	1·0000	−0·0266	−1·0777	0·8705	−0·5270	0·0000
0·0000	0·0000	−0·5915	−1·2655	−0·9204	−0·4904	1·0000

A multiple of the second pivotal row is subtracted from the row below it.

Step 8

1.0000	−0.4842	0.9247	3.3655	0.0000	1.0752	0.0000
0.0000	1.0000	0.0274	−0.9160	0.9076	−0.4765	0.0000
0.0000	0.0000	1.0000	2.7310	1.9863	1.0584	−1.6904

1.0000	−0.5161	0.8842	3.2736	0.0000	1.0526	0.0000
0.0000	1.0000	−0.0266	−1.0777	0.8705	−0.5270	0.0000
0.0000	0.0000	1.0000	1.4915	1.4074	0.6267	−2.1579

The last pivotal row of the interval matrix is divided by the pivot, 1.0000 being placed in the pivotal position.

Step 9

1.0000	−0.4842	0.0000	2.0467	−1.2444	0.5210	1.9955
0.0000	1.0000	0.0274	−0.9160	0.9076	−0.4765	0.0000
0.0000	0.0000	1.0000	2.7310	1.9863	1.0584	−1.6904

1.0000	−0.5161	0.0000	0.7482	−1.8368	0.0738	1.4946
0.0000	1.0000	−0.0266	−1.0777	0.8705	−0.5270	0.0000
0.0000	0.0000	1.0000	1.4915	1.4074	0.6267	−2.1579

The first step in the back-substitution procedure is to produce a zero in the first row of the third column of the coefficient matrix.

Step 10

1.0000	−0.4842	0.0000	2.0467	−1.2444	0.5210	1.9955
0.0000	1.0000	0.0000	−0.8433	0.9605	−0.4483	0.0591
0.0000	0.0000	1.0000	2.7310	1.9863	1.0584	−1.6904

1.0000	−0.5161	0.0000	0.7482	−1.8368	0.0738	1.4946
0.0000	1.0000	0.0000	−1.1526	0.8160	−0.5560	−0.0574
0.0000	0.0000	1.0000	1.4915	1.4074	0.6267	−2.1579

The second step in the back-substitution procedure is to produce a zero in the second row in the third column of the coefficient matrix.

Step 11

1.0000	0.0000	0.0000	1.6384	−0.7487	0.3039	2.0260
0.0000	1.0000	0.0000	−0.8433	0.9605	−0.4483	0.0591
0.0000	0.0000	1.0000	2.7310	1.9863	1.0584	−1.6904

1.0000	0.0000	0.0000	0.1533	−1.4417	−0.2131	1.4650
0.0000	1.0000	0.0000	−1.1526	0.8160	−0.5560	−0.0574
0.0000	0.0000	1.0000	1.4915	1.4074	0.6267	−2.1579

Finally, the back-substitution procedure produces a zero in the first row above the second pivot.

Step 11 gives us the endpoints for the solution vector and for the inverse, obtained by Gaussian elimination with partial pivoting using rounded interval arithmetic, these being our O.M. results.

As mentioned earlier, the four-figure results above represent more accurate results truncated to 4D.

The corresponding properly rounded 4D results give intervals which certainly contain the intervals for $\mathbf{x}^{\mathrm{I}}$ and $\mathbf{B}^{\mathrm{I}}$. We then have

$$\mathbf{x}^{\mathrm{I}} \subseteq \begin{bmatrix} [0 \cdot 1533, & 1 \cdot 6385] \\ [-1 \cdot 1527, & -0 \cdot 8433] \\ [1 \cdot 4915, & 2 \cdot 7311] \end{bmatrix}, \tag{9.33}$$

and

$$\mathbf{B}^{\mathrm{I}} \subseteq$$
$$\begin{bmatrix} [-1 \cdot 4418, & -0 \cdot 7487] & [-0 \cdot 2132, & 0 \cdot 3040] & [1 \cdot 4650, & 2 \cdot 0261] \\ [0 \cdot 8160, & 0 \cdot 9606] & [-0 \cdot 5561, & -0 \cdot 4483] & [-0 \cdot 0575, & 0 \cdot 0592] \\ [1 \cdot 4074, & 1 \cdot 9864] & [\ 0 \cdot 6267, & 1 \cdot 0585] & [-2 \cdot 1580, & -1 \cdot 6904] \end{bmatrix}$$
$$\tag{9.34}$$

Let us at this stage compare the intervals of uncertainty given by the O.M. results with those by other methods.

Of the methods previously dealt with it is Method I that leads to the widest intervals of uncertainty. For our example we have

$$w(U_i^{\mathrm{I}}) = 2\Delta x_i^{\mathrm{I}} = 2 \times 0 \cdot 3593 = 0 \cdot 7186, \qquad i = 1, 2, 3 \tag{9.35}$$

(see (3.25)).

But from (9.33) the widths of the intervals of uncertainty in x_1, x_2, and x_3 as obtained by O.M. are

$$\begin{aligned} 1 \cdot 6385 - 0 \cdot 1533 &= 1 \cdot 4852, \\ -0 \cdot 8433 - (-1 \cdot 1527) &= 0 \cdot 3094, \\ 2 \cdot 7311 - 1 \cdot 4915 &= 1 \cdot 2396, \end{aligned} \tag{9.36}$$

and

respectively.

On comparing (9.35) and (9.36) it is clear that two of the intervals by O.M. are larger than those by Method I.

And by (6.71) the widths of the intervals of uncertainty by Method IV for our example are

$$w(U_1^{\mathrm{IV}}) = 2\Delta x_1^{\mathrm{IV}} = 0 \cdot 3122, \qquad w(U_2^{\mathrm{IV}}) = 2\Delta x_2^{\mathrm{IV}} = 0 \cdot 1526, \tag{9.37}$$
$$w(U_3^{\mathrm{IV}}) = 2\Delta x_3^{\mathrm{IV}} = 0 \cdot 4800.$$

Clearly, the widths of the intervals of uncertainty by O.M. are at least twice the corresponding values by Method IV.

Further, since O.M. demands more computation than Methods I or IV, we have a clear indication that it is not a useful method.

Also, the field of application of O.M. is more restricted for our example than that of Method IV.

We have seen that in our example

$$\phi = 0 \cdot 08611$$

for an uncertainty of 0·01 (see (1.48)). But $\phi < 1$ is a sufficient condition for no critical ill-conditioning (see (2.46)). It follows by (1.36) that provided

$$\varepsilon < \frac{1}{0 \cdot 08611} \times 0 \cdot 01 = 0 \cdot 116 \qquad (9.38)$$

then $\phi < 1$ so that there is then no critical ill-conditioning. The uncertainties then exist and upper bounds of these can be obtained by Method IV (see (6.52)).

But experimentally it is found that O.M. fails for our example when

$$\varepsilon = 0 \cdot 072 \qquad (9.39)$$

because for $\varepsilon = 0 \cdot 072$ the third pivot includes zero in its interval. (The corresponding pivot for the case $\varepsilon = 0 \cdot 01$ is found in Step 7 of the computation scheme in the present section, the interval of this pivot being $[-0 \cdot 5915, -0 \cdot 4634]$.) The solution clearly cannot proceed further with zero in the interval of the third pivot for $\varepsilon = 0 \cdot 072$. For division by an interval containing zero is not defined. Or, looking at it in another way, the determinant of the coefficient matrix evaluated by O.M. includes zero in its interval value for $\varepsilon = 0 \cdot 072$, the value of a determinant apart from sign being given by the product of the pivots.

Thus, for $\varepsilon = 0 \cdot 072$, O.M. would indicate that the system of equations is critically ill-conditioned, while by (9.38) it is clear that critical ill-conditioning only occurs for values of ε much greater. (By (9.38) there is no critical ill-conditioning if $\varepsilon < 0 \cdot 116$.)

Clearly O.M. is not a suitable method for determining uncertainties.

In Section 9.4 we shall indicate the reasons why O.M. behaves so poorly. And in Section 9.5 it will be seen by way of example that if the system of equations is first suitably transformed then O.M. leads to intervals of satisfactory width containing the true intervals of uncertainty.

9.4 Dependence of Results on Precise Arithmetical Procedure

Suppose we are given that the value of a variable x lies in the interval $X = [1, 2]$, i.e.,

$$x \in X = [1, 2]. \tag{9.40}$$

Then, x has, of course, a unique value in any particular situation, say at a given instant of time. Suppose this value is 1·732.

And suppose we now multiply x by itself and then divide the result by x. This gives us a final result equal to x:

$$x = x.x/x = 1\cdot732 \times 1\cdot732/1\cdot732 = 1\cdot732.$$

It is thus easy to see that in general

$$\text{if} \quad x \in X = [1, 2] \quad \text{then} \quad x.x/x \in X = [1, 2]. \tag{9.41}$$

But suppose we determine the interval in which $x.x/x$ lies by carrying out the operations on intervals instead of using specific values of x. Then we have that

$$x.x/x \in [1, 2].[1, 2]/[1, 2] = [1, 4]/[1, 2] = [\tfrac{1}{2}, 4]. \tag{9.42}$$

Clearly, the final interval in (9.42) is an overestimate of the actual situation, the width of the interval in (9.42) being greater than the one in (9.41).

The reason for this is that the interval in (9.42) actually represents the result of

$$x.y/z \quad \text{for} \quad x \in [1, 2], \quad y \in [1, 2], \quad z \in [1, 2] \tag{9.43}$$

where x, y, and z can vary independently. For then for $x = 1$, $y = 1$, $z = 2$ we obtain the lower endpoint in (9.42), since $1 \times 1/2 = \tfrac{1}{2}$, and for $x = 2$, $y = 2$, $z = 1$ we obtain the upper endpoint in (9.42), since $2 \times 2/1 = 4$.

Thus, interval arithmetic may lead to intervals that are too wide if the same variable appears more than once in a computation. For interval arithmetic allows for different values in the various appearances of the variable whereas in fact the value of the variable is unique throughout the computation.

If we examine the Gaussian elimination process, it becomes clear that the coefficients enter more than once into the computation.

For example, solving the system of equations

$$\begin{aligned} a_{11}x_1 + a_{12}x_2 &= c_1 \\ a_{21}x_1 + a_{22}x_2 &= c_2 \end{aligned} \tag{9.44}$$

for x_1 gives

$$x_1 = \frac{a_{22}c_1 - a_{12}c_2}{a_{11}a_{22} - a_{12}a_{21}}. \tag{9.45}$$

Thus, the coefficients a_{22} and a_{12} appear more than once in the value of x_1.

Clearly, for an interval system of equations the Gaussian elimination process operating on intervals will in general lead to intervals that are too wide and that are not acceptable in our case.

But suppose that the given interval system of equations is a diagonal system of equations, for example

$$\begin{aligned} A_{11}x_1 \quad\quad &= C_1 \\ A_{22}x_2 &= C_2 \end{aligned} \tag{9.46}$$

where A_{11}, A_{22}, C_1, and C_2 are intervals (the interval coefficients A_{21} and A_{12} being the zero interval $[0, 0]$).
The solution of (9.46) then of course leads to

$$x_1 \in C_1/A_{11}, \quad\quad x_2 \in C_2/A_{22}. \tag{9.47}$$

But now the intervals for x_1 and x_2 are those of minimum width, i.e., they are the true intervals of uncertainty in the unknowns.

And in a strongly diagonal system of equations one may expect that Gaussian elimination on intervals will lead to intervals of uncertainty in the unknowns which are not much wider than the true ones.

We shall see in the next section that by suitably transforming the system of equations before solution to obtain a strongly diagonal system of equations, and then solving by O.M., leads to intervals for our example which are in fact less than those by Method IV.

9.5 Method VII

To overcome the difficulties of O.M., Hansen proposed multiplying the given system of equations

$$\mathbf{A}^I\mathbf{x} = \mathbf{c}^I \tag{9.48}$$

by $\mathbf{B} = \mathbf{A}^{-1}$ for any $\mathbf{A} \in \mathbf{A}^I$ (see reference 2).
In our treatment here, we choose $\mathbf{A}$ to correspond to the midpoint matrix of $\mathbf{A}^I$. Thus we choose

$$\mathbf{B} = \mathbf{A}^{-1} \tag{9.49}$$

where $\mathbf{A}$ is the approximate coefficient matrix of Chapter 1 (see (1.7) and (1.9)).

Then, carrying out the transformation

$$\mathbf{BA^I x = Bc^I} \qquad (9.50)$$

leads in practice to a strongly diagonal system of equations. This then may be expected to minimize the effects of the coefficients appearing more than once during the Gaussian elimination (see reference 2). And for our example this is certainly the case, as will be seen later in this section.

Let us now write

$$\mathbf{A^I = [A - \Delta A, A + \Delta A]}, \qquad \mathbf{c^I = [c - \Delta c, c + \Delta c]}, \quad (9.51)$$

so that $\mathbf{A}$ is the midpoint matrix of $\mathbf{A^I}$ and $\mathbf{c}$ is the midpoint vector of $\mathbf{c^I}$. Thus in terms of our previous notation

$$\mathbf{\Delta A} = (\varepsilon_{ij}), \qquad \mathbf{\Delta c} = (\varepsilon_i) \qquad (9.52)$$

(see (1.7)).

Then, multiplying

$$\mathbf{[A - \Delta A, A + \Delta A]x = [c - \Delta c, c + \Delta c]} \qquad (9.53)$$

by $\mathbf{B = A^{-1}}$ leads to

$$\mathbf{B[A - \Delta A, A + \Delta A]x = B[c - \Delta c, c + \Delta c]}. \qquad (9.54)$$

Now, by (9.31)

$$\begin{aligned}\mathbf{B[A - \Delta A, A + \Delta A]} &\subseteq \mathbf{[BA - |B|\,\Delta A, BA + |B|\,\Delta A]} \\ &= \mathbf{[I - |B|\,\Delta A, I + |B|\,\Delta A]}\end{aligned} \qquad (9.55)$$

and

$$\mathbf{B[c - \Delta c, c + \Delta c]} \subseteq \mathbf{[Bc - |B|\,\Delta c, Bc + |B|\,\Delta c]}.$$

But if we widen the interval coefficient matrix or the interval right-hand constant vector in (9.54) then the effect will be to widen the interval containing the unknowns.

Hence, by (9.54) and (9.55) it follows that the vector interval in $\mathbf{x}$ determined by

$$\mathbf{[I - |B|\,\Delta A, I + |B|\,\Delta A]x = [Bc - |B|\,\Delta c, Bc + |B|\,\Delta c]} \quad (9.56)$$

will contain the true intervals of uncertainty in the unknowns satisfying (9.48).

Thus, if X_1 is the solution set of (9.48) and X_2 is the solution set of (9.56) then

$$X_1 \subseteq X_2 \qquad (9.57)$$

where

$$X_1 = \{\mathbf{x} \mid \mathbf{Dx = e, D \in A^I, e \in c^I}\} \qquad (9.58)$$

and

$$X_2 = \{\mathbf{x} \mid \mathbf{Dx} = \mathbf{e}, \mathbf{D} \in [\mathbf{I} - |\mathbf{B}| \,\Delta\mathbf{A}, \mathbf{I} + |\mathbf{B}| \,\Delta\mathbf{A}],$$
$$\mathbf{e} \in [\mathbf{Bc} - |\mathbf{B}| \,\Delta\mathbf{c}, \mathbf{Bc} + |\mathbf{B}| \,\Delta\mathbf{c}]\}. \quad (9.59)$$

Our Method VII is then the solution by O.M. of the interval system of equations in (9.56), i.e., the determination by interval arithmetic of the solution set X_2 which contains the true solution set X_1 of the given approximate system of equations.

For our numerical example in (1.43) and (1.44), we have

$$\mathbf{A} = \begin{bmatrix} 0 \cdot 53 & 0 \cdot 86 & 0 \cdot 48 \\ 0 \cdot 94 & -0 \cdot 47 & 0 \cdot 85 \\ 0 \cdot 87 & 0 \cdot 55 & 0 \cdot 26 \end{bmatrix}, \quad \Delta\mathbf{A} = \begin{bmatrix} 0 \cdot 01 & 0 \cdot 01 & 0 \cdot 01 \\ 0 \cdot 01 & 0 \cdot 01 & 0 \cdot 01 \\ 0 \cdot 01 & 0 \cdot 01 & 0 \cdot 01 \end{bmatrix}, \quad (9.60)$$

$$\mathbf{c} = \begin{bmatrix} 0 \cdot 64 \\ 3 \cdot 12 \\ 0 \cdot 83 \end{bmatrix}, \quad \Delta\mathbf{c} = \begin{bmatrix} 0 \cdot 01 \\ 0 \cdot 01 \\ 0 \cdot 01 \end{bmatrix}, \quad (9.61)$$

and

$$\mathbf{B} = \mathbf{A}^{-1} = \begin{bmatrix} -1 \cdot 0574 & 0 \cdot 0724 & 1 \cdot 7153 \\ 0 \cdot 8878 & -0 \cdot 5017 & 0 \cdot 0013 \\ 1 \cdot 6603 & 0 \cdot 8189 & -1 \cdot 8963 \end{bmatrix}. \quad (9.62)$$

Hence, corresponding to Step 1 in the computation scheme in Section 9.3, the upper endpoint matrix consists of the columns of

$$\mathbf{I} + |\mathbf{B}| \,\Delta\mathbf{A}, \quad \mathbf{Bc} + |\mathbf{B}| \,\Delta\mathbf{c}, \quad \text{and} \quad \mathbf{BI} = \mathbf{B} \quad (9.63)$$

written side by side (see (9.56)). And the lower endpoint matrix is formed from the columns of

$$\mathbf{I} - |\mathbf{B}| \,\Delta\mathbf{A}, \quad \mathbf{Bc} - |\mathbf{B}| \,\Delta\mathbf{c}, \quad \text{and} \quad \mathbf{BI} = \mathbf{B}. \quad (9.64)$$

For our example in (1.43) and (1.44) (see (9.60), (9.61), and (9.62)) this then leads to

Step 1

1·0284	0·0284	0·0284	1·0014	−1·0574	0·0724	1·7153
0·0139	1·0139	0·0139	−0·9822	0·8877	−0·5017	0·0012
0·0437	0·0437	1·0437	2·0875	1·6602	0·8189	−1·8962

0·9715	−0·0284	−0·0284	0·9445	−1·0574	0·0724	1·7153
−0·0139	0·9860	−0·0139	−1·0100	0·8877	−0·5017	0·0012
−0·0437	−0·0437	0·9562	1·9999	1·6602	0·8189	−1·8962

The endpoints of the intervals of the transformed system of equations in (9.56). The coefficient matrix is now strongly diagonal.

The step corresponding to Step 2 of the computation scheme in Section 9.3 is no longer necessary, because there is no need now to interchange rows initially. And the steps corresponding to Steps 3 to 11 of the computation scheme in Section 9.3 follow below without explanation.

Step 3

1·0000	0·0292	0·0292	1·0307	−1·0281	0·0745	1·7655
0·0139	1·0139	0·0139	−0·9822	0·8877	−0·5017	0·0012
0·0437	0·0437	1·0437	2·0875	1·6602	0·8189	−1·8962

1·0000	−0·0292	−0·0292	0·9184	−1·0883	0·0704	1·6678
−0·0139	0·9860	−0·0139	−1·0100	0·8877	−0·5017	0·0012
−0·0437	−0·0437	0·9562	1·9999	1·6602	0·8189	−1·8962

Step 4

1·0000	0·0292	0·0292	1·0307	−1·0281	0·0745	1·7655
0·0000	1·0143	0·0143	−0·9679	0·9029	−0·5006	0·0258
0·0437	0·0437	1·0437	2·0875	1·6602	0·8189	−1·8962

1·0000	−0·0292	−0·0292	0·9184	−1·0883	0·0704	1·6678
0·0000	0·9856	−0·0143	−1·0243	0·8726	−0·5027	−0·0232
−0·0437	−0·0437	0·9562	1·9999	1·6602	0·8189	−1·8962

Step 5

1·0000	0·0292	0·0292	1·0307	−1·0281	0·0745	1·7655
0·0000	1·0143	0·0143	−0·9679	0·9029	−0·5006	0·0258
0·0000	0·0450	1·0450	2·1326	1·7079	0·8221	−1·8190

1·0000	−0·0292	−0·0292	0·9184	−1·0883	0·0704	1·6678
0·0000	0·9856	−0·0143	−1·0243	0·8726	−0·5027	−0·0232
0·0000	−0·0450	0·9549	1·9548	1·6126	0·8156	−1·9735

Step 6

1·0000	0·0292	0·0292	1·0307	−1·0281	0·0745	1·7655
0·0000	1·0000	0·0145	−0·9542	0·9160	−0·4936	0·0261
0·0000	0·0450	1·0450	2·1326	1·7079	0·8221	−1·8190

1·0000	−0·0292	−0·0292	0·9184	−1·0883	0·0704	1·6678
0·0000	1·0000	−0·0145	−1·0392	0·8603	−0·5100	−0·0236
0·0000	−0·0450	0·9549	1·9548	1·6126	0·8156	−1·9735

Step 7

1·0000	0·0292	0·0292	1·0307	−1·0281	0·0745	1·7655
0·0000	1·0000	0·0145	−0·9542	0·9160	−0·4936	0·0261
0·0000	0·0000	1·0456	2·1794	1·7491	0·8451	−1·8178

1·0000	−0·0292	−0·0292	0·9184	−1·0883	0·0704	1·6678
0·0000	1·0000	−0·0145	−1·0392	0·8603	−0·5100	−0·0236
0·0000	0·0000	0·9543	1·9080	1·5714	0·7926	−1·9746

Step 8

1·0000	0·0292	0·0292	1·0307	−1·0281	0·0745	1·7655
0·0000	1·0000	0·0145	−0·9542	0·9160	−0·4936	0·0261
0·0000	0·0000	1·0000	2·2837	1·8329	0·8856	−1·7384

1·0000	−0·0292	−0·0292	0·9184	−1·0883	0·0704	1·6678
0·0000	1·0000	−0·0145	−1·0392	0·8603	−0·5100	−0·0236
0·0000	0·0000	1·0000	1·8247	1·5027	0·7580	−2·0692

Step 9

1·0000	0·0292	0·0000	1·0976	−0·9744	0·1005	1·8261
0·0000	1·0000	0·0145	−0·9542	0·9160	−0·4936	0·0261
0·0000	0·0000	1·0000	2·2837	1·8329	0·8856	−1·7384

1·0000	−0·0292	0·0000	0·8515	−1·1420	0·0445	1·6072
0·0000	1·0000	−0·0145	−1·0392	0·8603	−0·5100	−0·0236
0·0000	0·0000	1·0000	1·8247	1·5027	0·7580	−2·0692

Step 10

1·0000	0·0292	0·0000	1·0976	−0·9744	0·1005	1·8261
0·0000	1·0000	0·0000	−0·9210	0·9426	−0·4807	0·0562
0·0000	0·0000	1·0000	2·2837	1·8329	0·8856	−1·7384

1·0000	−0·0292	0·0000	0·8515	−1·1420	0·0445	1·6072
0·0000	1·0000	0·0000	−1·0724	0·8337	−0·5229	−0·0536
0·0000	0·0000	1·0000	1·8247	1·5027	0·7580	−2·0692

Step 11

1·0000	0·0000	0·0000	1·1290	−0·9468	0·1158	1·8278
0·0000	1·0000	0·0000	−0·9210	0·9426	−0·4807	0·0562
0·0000	0·0000	1·0000	2·2837	1·8329	0·8856	−1·7384

1·0000	0·0000	0·0000	0·8201	−1·1696	0·0291	1·6056
0·0000	1·0000	0·0000	−1·0724	0·8337	−0·5229	−0·0536
0·0000	0·0000	1·0000	1·8247	1·5027	0·7580	−2·0692

The above results are the rounded interval arithmetic results obtained in a wordlength of eight significant figures and then truncated to four decimal places.

The properly rounded 4D results corresponding to Step 11 give intervals which certainly contain the true intervals of uncertainty. Thus,

$$
\begin{aligned}
x_1 &\in U_1^{\mathrm{VII}} = [0 \cdot 8201, \; 1 \cdot 1291], \\
x_2 &\in U_2^{\mathrm{VII}} = [-1 \cdot 0725, \; -0 \cdot 9210], \\
x_3 &\in U_3^{\mathrm{VII}} = [1 \cdot 8247, \; 2 \cdot 2838].
\end{aligned}
\tag{9.65}
$$

Hence, by (9.37) and (9.65) we have

$$
\begin{aligned}
w(U_1^{\mathrm{VII}}) &= 1\cdot1291 - 0\cdot8201 = 0\cdot3090 \le 0\cdot3122 = w(U_1^{\mathrm{IV}}), \\
w(U_2^{\mathrm{VII}}) &= -0\cdot9210 + 1\cdot0725 = 0\cdot1515 \le 0\cdot1526 = w(U)_2^{\mathrm{IV}}, \quad (9.66) \\
w(U_3^{\mathrm{VII}}) &= 2\cdot2838 - 1\cdot8247 = 0\cdot4591 \le 0\cdot4800 = w(U_3^{\mathrm{IV}}).
\end{aligned}
$$

Summarizing, then, we may say that:

> Method VII always leads to intervals of uncertainty containing the true intervals of uncertainty and can lead to intervals of smaller width than those by Method IV (see (9.66)).   (9.67)

In the next section we give some experimental results of comparing Methods VII and IV. And in Section 9.7 we derive upper bounds of the uncertainties by interval arithmetic considerations which are the same as those by Method IV. The reader may, however, at this stage proceed directly to Section 9.8 in which we describe our Method VIII.

9.6   Comparison of Methods VII and IV

In this section we give some experimental results for our example and for another example frequently quoted in the literature.

First, we consider the field of application of Method VII.

Let us then recall that we saw in Section 9.3 (see (9.39)) that O.M. failed for our example for $\varepsilon = 0\cdot072$ corresponding to $\phi = 0\cdot62$. And, although Method VII may be expected to fail at a higher value of ϕ, it is interesting to see how close to unity this value is, since the condition of application of its rival, Method IV, is $\phi < 1$ (for the case when the uncertainties in the coefficients are all equal).

In fact, Method VII succeeds in giving uncertainties for our example in (1.43) for ε corresponding to $\phi = 1 - 10^{-10}$. This is certainly close enough to 1.

Now let us recall that the condition $\phi < 1$ is in general only a sufficient condition for no critical ill-conditioning. Thus it might occur that Method VII succeeds for $\phi > 1$, a region in which Method IV is not applicable. But Method VII fails for our example for ε corresponding to $\phi = 1 + 10^{-10}$.

Clearly, then, for our example the field of application of Methods IV and VII are the same for all practical purposes.

And in the frequently quoted example in the literature (see reference 3)

$$\begin{aligned}
4{\cdot}33x_1 - 1{\cdot}12x_2 - 1{\cdot}08x_3 + 1{\cdot}14x_4 &= 3{\cdot}52 \\
-1{\cdot}12x_1 + 4{\cdot}33x_2 + 0{\cdot}24x_3 - 1{\cdot}22x_4 &= 1{\cdot}57 \\
-1{\cdot}08x_1 + 0{\cdot}24x_2 + 7{\cdot}21x_3 - 3{\cdot}22x_4 &= 0{\cdot}54 \\
1{\cdot}14x_1 - 1{\cdot}22x_2 - 3{\cdot}22x_3 + 5{\cdot}43x_4 &= -1{\cdot}09
\end{aligned} \tag{9.68}$$

Method VII also succeeds in determining finite uncertainties for ε corresponding to $\phi = 1 - 10^{-10}$, but fails for ε corresponding to $\phi = 1 + 10^{-10}$.

Next, let us compare the widths of the intervals of uncertainty by Methods VII and IV.

For our example in (1.43) for $\varepsilon = 0{\cdot}01$ for which $\phi = 0{\cdot}08611$ the ratios of the widths of the intervals of uncertainty by Methods VII to IV are

$$\begin{aligned}
w(U_1^{\mathrm{VII}})/w(U_1^{\mathrm{IV}}) &= 0{\cdot}9898 \\
w(U_2^{\mathrm{VII}})/w(U_2^{\mathrm{IV}}) &= 0{\cdot}9920 \\
w(U_3^{\mathrm{VII}})/w(U_3^{\mathrm{IV}}) &= 0{\cdot}9563.
\end{aligned} \tag{9.69}$$

It may at this stage be interesting to compare the widths of the intervals of uncertainty by Methods VII and IV for various values of ϕ for our example (see Table 9.1).

TABLE 9.1   Ratios of $w(U_i^{\mathrm{VII}})$ to $w(U_i^{\mathrm{IV}})$ for (1.43)

ϕ	$w(U_1^{\mathrm{VII}})/w(U_1^{\mathrm{IV}})$	$w(U_2^{\mathrm{VII}})/w(U_2^{\mathrm{IV}})$	$w(U_3^{\mathrm{VII}})/w(U_3^{\mathrm{IV}})$
0·1	0·9882	0·9907	0·9492
0·2	0·9791	0·9824	0·8987
0·3	0·9724	0·9751	0·8483
0·4	0·9682	0·9693	0·7980
0·5	0·9665	0·9653	0·7479
0·6	0·9675	0·9637	0·6980
0·7	0·9711	0·9652	0·6483
0·8	0·9776	0·9708	0·5987
0·9	0·9871	0·9817	0·5923

It would appear from Table 9.1 that for ϕ small the results by Methods VII and IV are nearly the same. Indeed, for $\phi = 10^{-10}$ the ratios $w(U_i^{\mathrm{VII}})/w(U_i^{\mathrm{IV}})$ for $i = 1, 2, 3$ are

$$0{\cdot}999999999986, \quad 0{\cdot}999999999990, \quad 0{\cdot}999999999949. \tag{9.70}$$

Further, it would appear from Table 9.1 that the first two ratios approach unity as ϕ approaches unity. In fact, for $\phi = 1 - 10^{-10} = 0{\cdot}9999999999$ the three ratios corresponding to those in (9.70) are

$$0{\cdot}999999999985, \quad 0{\cdot}999999999977, \quad 0{\cdot}5923. \tag{9.71}$$

And results similar to those for our example hold for the system of equations in (9.68).

Thus, Method VII has the same field of application for the examples in (1.43) and (9.68) as Method IV. But the uncertainties by Method VII are better than those by Method IV for the examples.

Now in the next section we obtain theoretical upper bounds of the uncertainties by interval arithmetic considerations after carrying out Hansen's transformation; these are in fact found to be equal to the uncertainties by Method IV.

9.7 Upper Bounds of Uncertainties by Interval Arithmetic

In this section we determine, by interval arithmetic considerations, expressions for the endpoints of the intervals of uncertainty such that the intervals so obtained contain the true intervals of uncertainty. (Using interval arithmetic ensures this property.) This leads to upper bounds of the uncertainties which are the same as those by Method IV.

First, we determine in (9.83) by interval arithmetic considerations an interval containing the interval inverse of

$$[\mathbf{I} - \mathbf{C}, \mathbf{I} + \mathbf{C}] \tag{9.72}$$

where $\mathbf{C}$ is the ill-conditioning matrix, assuming that

$$\rho(\mathbf{C}) < 1. \tag{9.73}$$

Now, by (9.27) we have

$$[\mathbf{I} - \mathbf{C}, \mathbf{I} + \mathbf{C}]^{-1} = \{(\mathbf{I} - \mathbf{E})^{-1} \mid (\mathbf{I} - \mathbf{E}) \in [\mathbf{I} - \mathbf{C}, \mathbf{I} + \mathbf{C}]\} \tag{9.74}$$

provided

$$\text{no } \det(\mathbf{I} - \mathbf{E}) = 0. \tag{9.75}$$

But clearly

$$\begin{aligned}
\{(\mathbf{I} - \mathbf{E})^{-1} \mid (\mathbf{I} - \mathbf{E}) \in [\mathbf{I} - \mathbf{C}, \mathbf{I} + \mathbf{C}]\} \\
= \{(\mathbf{I} - \mathbf{E})^{-1} \mid -\mathbf{E} \in [-\mathbf{C}, \mathbf{C}]\} \\
= \{(\mathbf{I} - \mathbf{E})^{-1} \mid \mathbf{E} \in [-\mathbf{C}, \mathbf{C}]\}.
\end{aligned}$$

Hence, let us rewrite (9.74) and (9.75) as

$$[\mathbf{I} - \mathbf{C}, \mathbf{I} + \mathbf{C}]^{-1} = \{(\mathbf{I} - \mathbf{E})^{-1} \mid \mathbf{E} \in [-\mathbf{C}, \mathbf{C}]\}, \tag{9.76}$$

provided

$$\det(\mathbf{I} - \mathbf{E}) \neq 0 \quad \text{for} \quad \mathbf{E} \in [-\mathbf{C}, \mathbf{C}]. \tag{9.77}$$

Now, this condition is satisfied by (6.35) if

$$\rho(\mathbf{E}) < 1 \quad \text{for} \quad \mathbf{E} \in [-\mathbf{C}, \mathbf{C}]. \tag{9.78}$$

But clearly, $|\mathbf{E}| \leq \mathbf{C}$ for $\mathbf{E} \in [-\mathbf{C}, \mathbf{C}]$, so that we have by (2.34) that

$$\rho(\mathbf{E}) < 1 \quad \text{if} \quad \rho(\mathbf{C}) < 1 \quad \text{for} \quad \mathbf{E} \in [-\mathbf{C}, \mathbf{C}] \tag{9.79}$$

Hence, by (9.76), (9.77), (9.78), and (9.79) it follows that

$$[\mathbf{I} - \mathbf{C}, \mathbf{I} + \mathbf{C}]^{-1} \quad \text{exists if} \quad \rho(\mathbf{C}) < 1 \tag{9.80}$$

and more particularly that

$$[\mathbf{I} - \mathbf{C}, \mathbf{I} + \mathbf{C}]^{-1} = \{(\mathbf{I} - \mathbf{E})^{-1} \mid \mathbf{E} \in [-\mathbf{C}, \mathbf{C}]\} \quad \text{if} \quad \rho(\mathbf{C}) < 1. \tag{9.81}$$

It then follows by (9.73), (9.79), (2.27), and (2.23) that

$$[\mathbf{I} - \mathbf{C}, \mathbf{I} + \mathbf{C}]^{-1} = \{\mathbf{I} + \mathbf{E} + \mathbf{E}^2 + \mathbf{E}^3 + \ldots \mid \mathbf{E} \in [-\mathbf{C}, \mathbf{C}]\}. \tag{9.82}$$

Since the ill-conditioning matrix $\mathbf{C} \geq \mathbf{O}$, it now clearly follows from (9.82) that

$$\begin{aligned}
[\mathbf{I} - \mathbf{C}, \mathbf{I} + \mathbf{C}]^{-1} \\
&\subseteq [\mathbf{I} - \mathbf{C} - \mathbf{C}^2 - \mathbf{C}^3 - \ldots, \mathbf{I} + \mathbf{C} + \mathbf{C}^2 + \mathbf{C}^3 + \ldots] \\
&= [\mathbf{I} - (\mathbf{I} + \mathbf{C} + \mathbf{C}^2 + \ldots)\mathbf{C}, \mathbf{I} + (\mathbf{I} + \mathbf{C} + \mathbf{C}^2 + \ldots)\mathbf{C}] \\
&= [\mathbf{I} - (\mathbf{I} - \mathbf{C})^{-1} \mathbf{C}, \mathbf{I} + (\mathbf{I} - \mathbf{C})^{-1} \mathbf{C}]
\end{aligned}$$

by (2.27) and (9.73).

Thus,

$$\begin{aligned}
[\mathbf{I} - \mathbf{C}, \mathbf{I} + \mathbf{C}]^{-1} \subseteq [\mathbf{I} - (\mathbf{I} - \mathbf{C})^{-1} \mathbf{C}, \mathbf{I} + (\mathbf{I} - \mathbf{C})^{-1} \mathbf{C}], \\
\text{provided } \rho(\mathbf{C}) < 1.
\end{aligned} \tag{9.83}$$

We now proceed to determine intervals containing the true intervals of uncertainty assuming that Hansen's transformation is carried out.

Now, by (9.56), (9.57), and (1.37) the intervals of uncertainty determined by

$$[\mathbf{I} - \mathbf{C}, \mathbf{I} + \mathbf{C}]\mathbf{x} = [\mathbf{Bc} - |\mathbf{B}| \, \Delta\mathbf{c}, \mathbf{Bc} + |\mathbf{B}| \, \Delta\mathbf{c}] \tag{9.84}$$

will certainly contain the true intervals of uncertainty in the unknowns.

Hence, assuming that

$$\rho(\mathbf{C}) < 1 \tag{9.85}$$

we have by (9.80) that

$$\begin{aligned}
\mathbf{x} &\in [\mathbf{I} - \mathbf{C}, \mathbf{I} + \mathbf{C}]^{-1} [\mathbf{Bc} - |\mathbf{B}| \, \Delta\mathbf{c}, \mathbf{Bc} + |\mathbf{B}| \, \Delta\mathbf{c}] \tag{9.86} \\
&\subseteq [\mathbf{I} - (\mathbf{I} - \mathbf{C})^{-1} \mathbf{C}, \mathbf{I} + (\mathbf{I} - \mathbf{C})^{-1} \mathbf{C}] [\mathbf{Bc} - |\mathbf{B}| \, \Delta\mathbf{c}, \mathbf{Bc} + |\mathbf{B}| \, \Delta\mathbf{c}], \\
&\qquad \text{by (9.83),}
\end{aligned}$$

$$\begin{aligned}
&\subseteq [\mathbf{Bc} - |\mathbf{B}| \, \Delta\mathbf{c} - (\mathbf{I} - \mathbf{C})^{-1} \mathbf{C} \, |\mathbf{Bc}| - (\mathbf{I} - \mathbf{C})^{-1} \mathbf{C} \, |\mathbf{B}| \, \Delta\mathbf{c}, \\
&\quad\; \mathbf{Bc} + |\mathbf{B}| \, \Delta\mathbf{c} + (\mathbf{I} - \mathbf{C})^{-1} \mathbf{C} \, |\mathbf{Bc}| + (\mathbf{I} - \mathbf{C})^{-1} \mathbf{C} \, |\mathbf{B}| \, \Delta\mathbf{c}].
\end{aligned} \tag{9.87}$$

Now, since the midpoint of the interval in (9.87) is $\mathbf{Bc}$, the solution of $\mathbf{Ax} = \mathbf{c}$, it follows for the uncertainty $\Delta\mathbf{x}$ in $\mathbf{x}$ that

$$
\begin{aligned}
\Delta\mathbf{x} \leq\ & |\mathbf{B}|\,\Delta\mathbf{c} + (\mathbf{I} - \mathbf{C})^{-1}\,\mathbf{C}\,|\mathbf{Bc}| + (\mathbf{I} - \mathbf{C})^{-1}\,\mathbf{C}\,|\mathbf{B}|\,\Delta\mathbf{c} \\
=\ & |\mathbf{B}|\,\Delta\mathbf{c} + (\mathbf{I} - \mathbf{C})^{-1}\,\mathbf{C}\,|\mathbf{x}| + (\mathbf{I} - \mathbf{C})^{-1}\,\mathbf{C}\,|\mathbf{B}|\,\Delta\mathbf{c},
\end{aligned}
\tag{9.88}
$$

$$
\text{since } \mathbf{Ax} = \mathbf{c}, \quad \text{so that } \mathbf{x} = \mathbf{Bc},
$$

$$
\begin{aligned}
=\ & (\mathbf{I} - \mathbf{C})^{-1}\,\mathbf{C}\,|\mathbf{x}| + (\mathbf{I} + (\mathbf{I} + \mathbf{C} + \mathbf{C}^2 + \ldots)\mathbf{C})\,|\mathbf{B}|\,\Delta\mathbf{c} \\
=\ & (\mathbf{I} - \mathbf{C})^{-1}\,\mathbf{C}\,|\mathbf{x}| + (\mathbf{I} + \mathbf{C} + \mathbf{C}^2 + \mathbf{C}^3 + \ldots)\,|\mathbf{B}|\,\Delta\mathbf{c} \\
=\ & (\mathbf{I} - \mathbf{C})^{-1}\,\mathbf{C}\,|\mathbf{x}| + (\mathbf{I} - \mathbf{C})^{-1}\,|\mathbf{B}|\,\Delta\mathbf{c}.
\end{aligned}
$$

Thus

$$
\begin{aligned}
\Delta\mathbf{x} \leq\ & (\mathbf{I} - \mathbf{C})^{-1}\,(\mathbf{C}\,|\mathbf{x}| + |\mathbf{B}|\,\Delta\mathbf{c}) \\
=\ & (\mathbf{I} - \mathbf{C})^{-1}\,\Delta\mathbf{x}^{\mathrm{II}}, \quad \text{by (4.23)}, \\
=\ & \Delta\mathbf{x}^{\mathrm{IV}}, \quad \text{by (6.16)}.
\end{aligned}
\tag{9.89}
$$

Hence:

Upper bounds for the uncertainties equal to those by Method IV can be obtained by interval arithmetic considerations, provided $\rho(\mathbf{C}) < 1$, this being the same condition as in Method IV. (9.90)

Thus, commencing with (9.84) in which Hansen's transformation is incorporated, we have derived theoretical upper bounds of the uncertainties equal to those given by Method IV.

But one would not like to claim that the numerical results by Method VII would always be better than those by Method IV as in the examples in Section 9.6. For, while in an interval arithmetic computation the results depend on the precise arithmetical steps, we made no use of the particular arithmetical procedure (Gaussian elimination by partial pivoting) in deriving (9.90).

9.8 Method VIII

We recall that in Method VI (Chapter 8) we combine Methods III and IV, Method IV being initially used to find the intervals of uncertainty in the unknowns and in the elements of the inverse and then whenever necessary to find the uncertainties in the approximate systems of equations being solved.

Now, Method VIII is a combination of Methods III and VII, Method VII being used in Method VIII whenever Method IV would be used in Method VI.

Thus, in Method VIII the intervals of uncertainty in the unknowns and in the elements of the inverse are initially found by Method VII. And in those cases where Method VIII does not reduce to Method III the intervals of uncertainty in the approximate system of equations are found by Method VII.

Let us then briefly outline Method VIII.

First, let us denote the interval of uncertainty by Method VII in the element b_{ki} of the inverse by

$$U_{ki}^{VII}, \tag{9.91}$$

the double suffix to indicate that we are dealing with an interval of uncertainty in an element of the inverse. And the interval of uncertainty by Method VII in the unknown x_i is of course denoted by U_i^{VII}.

Then, corresponding to (8.10) (see also (8.38)), the following is a set of sufficient conditions to ensure that the interval of uncertainty in x_k given by Method III is the true interval of uncertainty

$$0 \notin \text{int}(U_i^{VII}), \qquad 0 \notin \text{int}(U_{ki}^{VII}), \qquad i = 1, 2, \ldots, n. \tag{9.92}$$

But suppose that the conditions in (9.92) are not wholly satisfied for some value of k.

Then, corresponding to (8.12) to (8.15), the coefficients, constants, and uncertainties of the approximate system of equations which Method VIII requires to be solved by Method VII to determine the maximum increase in x_k are obtained from the following specifications:

$$\begin{aligned} \delta a_{ij} &= \varepsilon_{ij} \, \text{sign}(-x_j b_{ki}) \text{ with zero uncertainty in} \\ (a_{ij} &+ \delta a_{ij}) \quad \text{if} \quad 0 \notin \text{int}(U_j^{VII}) \quad \text{and} \quad 0 \notin \text{int}(U_{ki}^{VII}) \end{aligned} \tag{9.93}$$

but

$$\delta a_{ij} = 0 \text{ with uncertainty } \varepsilon_{ij} \text{ in } a_{ij} \text{ if } 0 \in \text{int}(U_j^{VII})$$
$$\text{or } 0 \in \text{int}(U_{ki}^{VII}), \tag{9.94}$$

and similarly

$$\delta c_i = \varepsilon_i \, \text{sign}(b_{ki}) \text{ with zero uncertainty in } (c_i + \delta c_i)$$
$$\text{if } 0 \notin \text{int}(U_{ki}^{VII}), \tag{9.95}$$

but

$$\delta c_i = 0 \text{ with an uncertainty } \varepsilon_i \text{ in } c_i \text{ if } 0 \in \text{int}(U_{ki}^{VII}). \tag{9.96}$$

With the changes and uncertainties as specified above, we then have to solve by Method VII a system of equations similar to that given in (8.16).

Suppose we obtain the interval $[r_k, s_k]$ for the kth unknown. Then this interval contains the upper endpoint of the interval of uncertainty in x_k of the original system of equations, i.e.,

$$x_k + d_k \in [r_k, s_k], \quad \text{i.e.,} \quad r_k \leq x_k + d_k \leq s_k \quad (9.97)$$

(see (1.11) and also (8.19)).

And let $[p_k, q_k]$ be the interval given by Method VII when determining the lower endpoint of the interval of uncertainty in x_k, so that

$$x_k - e_k \in [p_k, q_k], \quad \text{i.e.,} \quad p_k \leq x_k - e_k \leq q_k \quad (9.98)$$

(see (8.20), (1.11), and also (8.21)).

Then, corresponding to (8.23) and (8.24), we define the intervals U_k^{VIIIB} and U_k^{VIII} by

$$U_k^{\text{VIIIB}} = [q_k, r_k], \qquad U_k^{\text{VIII}} = [p_k, s_k], \quad (9.99)$$

U_k^{VIII} being referred to as the kth *interval of uncertainty by Method VIII*.

And, corresponding to (8.11), (8.22), (8.28), (8.29), and (8.30) we therefore now have

$$U_k^{\text{VIII}} = U_k^{\text{III}} = U_k \quad \text{if} \quad 0 \notin \text{int}(U_i^{\text{VII}}) \quad \text{and} \quad 0 \notin \text{int}(U_{ki}^{\text{VII}}), \quad i = 1, 2, \ldots, n, \quad (9.100)$$

$$U_k^{\text{VIIIB}} \subseteq U_k \subseteq U_k^{\text{VIII}} \quad \text{if (9.92) is not wholly satisfied} \quad (9.101)$$

$$U_k^{\text{VIII}} = U_k^{\text{VII}} \quad \text{if} \quad 0 \in \text{int}(U_{ki}^{\text{VII}}), \quad i = 1, 2, \ldots, n, \quad (9.102)$$

$$U_k^{\text{VIII}} \subseteq U_k^{\text{VII}} \quad (9.103)$$

$$U_k^{\text{VIIIB}} \subseteq U_k \subseteq U_k^{\text{VIII}} \subseteq U_k^{\text{VII}}, \quad k = 1, 2, \ldots, n. \quad (9.104)$$

Improved Version of Method VIII

The improved version of Method VIII bears the same relation to Method VIII as does the improved version of Method VI in Section 8.3 to Method VI.

In particular, corresponding to (8.40) to (8.43), the following are the specifications for the improved version of Method VIII in place of those in (9.93) to (9.96) for the ordinary version of Method VIII:

Put $\delta a_{ij} = \varepsilon_{ij} \, \text{sign}(-x_j b_{ki})$ with zero uncertainty in $(a_{ij} + \delta a_{ij})$ if both (a) and (b) hold:

(a) $0 \notin \text{int}(U_j^{\text{VII}})$ or $0 \notin \text{int}(U_j^{\text{VIII}})$ (9.105)

(b) $0 \notin \text{int}(U_{ki}^{\text{VII}})$ or $0 \notin \text{int}(U_{ki}^{\text{VIII}})$, but

put $\delta a_{ij} = 0$ with an uncertainty ε_{ij} in a_{ij} if either (a) or (b) in (9.105) does not hold, $\qquad$ (9.106)

and similarly

put $\delta c_i = \varepsilon_i \operatorname{sign}(b_{ki})$ with zero uncertainty in $(c_i + \delta c_i)$ if $\quad 0 \notin \operatorname{int}(U_{ki}^{\text{VII}}) \quad$ or $\quad 0 \notin \operatorname{int}(U_{ki}^{\text{VIII}}) \qquad$ (9.107)

but

put $\delta c_i = 0$ with an uncertainty ε_i in c_i if neither condition in (9.107) holds. $\qquad$ (9.108)

Example

For our example in (1.43) and (1.44) we have by (9.65) that

$$0 \notin \operatorname{int}(U_i^{\text{VII}}), \qquad i = 1, 2, 3. \qquad (9.109)$$

And by Step 11 in Section 9.5 we have that

$$0 \notin \operatorname{int}(U_{ki}^{\text{VII}}), \qquad k, i = 1, 2, 3 \qquad (9.110)$$

except that

$$0 \in \operatorname{int}(U_{23}^{\text{VII}}) = \operatorname{int}([-0{\cdot}0537, 0{\cdot}0563]). \qquad (9.111)$$

It therefore follows by (9.100) that for $k = 1$ and $k = 3$ we have

$$U_1^{\text{VII}} = U_1^{\text{III}} = U_1 \qquad \text{and} \qquad U_3^{\text{VII}} = U_3^{\text{III}} = U_3, \qquad (9.112)$$

respectively (see (8.50) and also (5.53)).

But for $k = 2$ we must determine the intervals in (9.99). These are found to be

$$U_2^{\text{VIIIB}} = [-1{\cdot}06679309, -0{\cdot}92770958]$$
$$U_2^{\text{VIII}} = [-1{\cdot}06719825, -0{\cdot}92758329]. \qquad (9.113)$$

Since by (9.101)

$$U_2^{\text{VIIIB}} \subseteq U_2 \subseteq U_2^{\text{VIII}}$$

the interval of uncertainty in x_2 is closely defined, the difference in widths between the intervals U_2^{VIII} and U_2^{VIIIB} being

$$w(U_2^{\text{VIII}}) - w(U_2^{\text{VIIIB}}) = 0{\cdot}13961495 - 0{\cdot}13908351$$
$$= 0{\cdot}00053144, \qquad (9.114)$$

all values given being obtained by truncating more accurate ones.

Let us now compare the results for our example by Methods VI and VIII.

By (8.64) and (9.113) U_2^{VI} and U_2^{VIII} are equal to 8D. But using more accurate values we have that

$$w(U_2^{VI}) - w(U_2^{VIII}) = 0{\cdot}13961495877 - 0{\cdot}13961495788$$
$$= 0{\cdot}00000000089. \qquad (9.115)$$

Hence, the result by Method VIII for the interval of uncertainty in x_2 is marginally better than by Method VI. For it is the width of the interval containing the true interval of uncertainty that is perhaps more important than how sharply the endpoints are determined (compare (8.65) and (9.114)).

Further, the volume of computation in Method VIII is of the same order of magnitude as in Method VI, and the same remarks also hold in comparing the volume of computation in Methods VII and IV.

Concluding this chapter, then, we should say that available evidence indicates that the interval arithmetic Methods VII and VIII are marginally to be preferred to Methods IV and VI.

In the next two chapters we shall see that at the cost of applying linear programming techniques it is possible to determine the exact intervals of uncertainty for non-critically ill-conditioned systems of equations.

References

1. HANSEN, E. 'Interval arithmetic in matrix computations', Pt I, SIAM *J. Numer. Anal.*, **2**, 308–20 (1965).
2. HANSEN, E. and SMITH, R. 'Interval arithmetic in matrix computations', Pt II, SIAM *J. Numer. Anal.*, **4**, 1–9 (1967).
3. HANSEN, E. 'On the Solution of Linear Algebraic Equations with Interval Coefficients', *Linear Algebra and Its Applications*, **2**, 153–65 (1969).
4. MOORE, R. E. *Interval Analysis*, Prentice-Hall, Englewood Cliffs, N.J. (1966).

CHAPTER 10

The Set of all Solutions

10.1 Introduction

We are considering an approximate system of equations, in which the true coefficient matrix $\mathbf{A}^*$ and the true right-hand column of constants $\mathbf{c}^*$ are known no more precisely than that given by

$$\mathbf{A}^* \in [\mathbf{A} - \Delta\mathbf{A}, \mathbf{A} + \Delta\mathbf{A}] = \mathbf{A}^I, \text{ say,}$$
$$\mathbf{c}^* \in [\mathbf{c} - \Delta\mathbf{c}, \mathbf{c} + \Delta\mathbf{c}] = \mathbf{c}^I, \text{ say,} \tag{10.1}$$

where

$$\Delta\mathbf{A} = (\varepsilon_{ij}) \quad \text{and} \quad \Delta\mathbf{c} = (\varepsilon_i) \tag{10.2}$$

(see (1.7)).

Then $\mathbf{x}$ is an *admissible solution* of the approximate system of equations if for some $\mathbf{A}^*$ and some $\mathbf{c}^*$

$$\mathbf{A}^*\mathbf{x} = \mathbf{c}^* \quad \text{where} \quad \mathbf{A}^* \in \mathbf{A}^I, \mathbf{c}^* \in \mathbf{c}^I. \tag{10.3}$$

And the totality of admissible solutions constitutes the *set of all solutions* X, i.e., the *solution set* X is given by

$$X = \{\mathbf{x} \mid \mathbf{A}^*\mathbf{x} = \mathbf{c}^*, \mathbf{A}^* \in \mathbf{A}^I, \mathbf{c}^* \in \mathbf{c}^I\}. \tag{10.4}$$

Now, in Section 10.2 we derive the necessary and sufficient conditions for a solution $\mathbf{x}$ to be an admissible solution. The solution set X can then be considered as the set of all $\mathbf{x}$ satisfying the necessary and sufficient conditions for an admissible solution.

The properties of X that are of interest to us are clearly the minimum and maximum values of each component of all $\mathbf{x}$ in X. And equivalent to the definition in (1.12) the intervals of uncertainty U_i may be defined by

$$U_i = [(x_i)_{\min}, (x_i)_{\max}], \qquad i = 1, 2, \ldots, n, \tag{10.5}$$

such that

$$(x_i)_{\min} = \min(x_i), \qquad \text{for all } \mathbf{x} \in X,$$
$$(x_i)_{\max} = \max(x_i), \qquad \text{for all } \mathbf{x} \in X. \tag{10.6}$$

Now, Methods I to VIII are concerned with finding the intervals of uncertainty. Most of these methods ensure that the estimates of the intervals of uncertainty always contain the true intervals of uncertainty and some give bounds on the endpoints of the intervals of uncertainty. And Methods VI and VIII, when they reduce to Method III, give the true intervals of uncertainty.

But the true intervals of uncertainty can always be found by using the necessary and sufficient conditions for admissible solutions and applying linear programming techniques (Chapter 11). And it will be seen by way of an example that this approach succeeds when all the other methods fail to give finite intervals of uncertainty. However, the volume of computation in the linear programming approach can in certain circumstances be very large.

The necessary and sufficient conditions for admissible solutions can also be used to obtain a clearer picture in certain situations of the solution set than that given only by the intervals of uncertainty. For example, if in fact some of the components x_i of a solution vector become known then the intervals in which the other unknowns may lie can become restricted within their intervals of uncertainty (Section 10.7). Thus, using the known values for some of the components of a solution vector, the widths of the intervals in which the other unknowns may lie can become smaller than those of their intervals of uncertainty, linear programming being used to obtain these smaller intervals.

Then, in Sections 10.8 and 10.9, we derive certain properties concerning the continuity and boundedness of the solution set. And in Section 10.10 we derive a sufficient condition for no critical ill-conditioning which may succeed in indicating no critical ill-conditioning when the other sufficient conditions for no critical ill-conditioning fail to do so. And it will be seen that when the condition is applicable then testing it is merely a by-product of obtaining the intervals of uncertainty from the conditions for admissible solutions.

10.2   The Set of All Solutions

In this section, we find the necessary and sufficient conditions for a vector $\mathbf{x}$ to be an admissible solution of

$$\mathbf{A}^*\mathbf{x} = \mathbf{c}^* \tag{10.7}$$

subject to (10.1) and (10.2). And the set of all solutions can then be regarded as specified by the necessary and sufficient conditions for admissible solutions.

Let us now suppose that $\mathbf{x}$ is an admissible solution and let us write (10.7) as

$$(\mathbf{A} + \delta\mathbf{A})\mathbf{x} = \mathbf{c} + \delta\mathbf{c} \tag{10.8}$$

where by (10.1) and (10.2)

$$(|\delta a_{ij}|) = |\delta \mathbf{A}| \leq \Delta \mathbf{A} = (\varepsilon_{ij}) \quad \text{and} \quad (|\delta c_i|) = |\delta \mathbf{c}| \leq \Delta \mathbf{c} = (\varepsilon_i).$$
(10.9)

Now, on multiplying (10.8) out we have

$$(\delta \mathbf{A})\mathbf{x} - \delta \mathbf{c} = \mathbf{c} - \mathbf{Ax}, \tag{10.10}$$

i.e.,

$$\sum_{j=1}^{n} \delta a_{ij}x_j - \delta c_i = c_i - \sum_{j=1}^{n} a_{ij}x_j, \qquad i = 1, 2, \ldots, n. \tag{10.11}$$

But subject to (10.9) the left-hand sides of (10.11)

$$\sum_{j=1}^{n} \delta a_{ij}x_j - \delta c_i, \qquad i = 1, 2, \ldots, n, \tag{10.12}$$

certainly do not lie outside the intervals

$$\left[-\sum_{j=1}^{n} \varepsilon_{ij} |x_j| - \varepsilon_i, \sum_{j=1}^{n} \varepsilon_{ij} |x_j| + \varepsilon_i \right], \qquad i = 1, 2, \ldots, n. \tag{10.13}$$

Hence, a necessary condition for $\mathbf{x}$ to be an admissible solution is that the right-hand sides of (10.11)

$$c_i - \sum_{j=1}^{n} a_{ij}x_j, \qquad i = 1, 2, \ldots, n, \tag{10.14}$$

do not lie outside the intervals in (10.13), i.e., that

$$\sum_{j=1}^{n} \varepsilon_{ij} |x_j| + \varepsilon_i \geq c_i - \sum_{j=1}^{n} a_{ij}x_j, \qquad i = 1, 2, \ldots, n, \tag{10.15}$$

and

$$-\sum_{j=1}^{n} \varepsilon_{ij} |x_j| - \varepsilon_i \leq c_i - \sum_{j=1}^{n} a_{ij}x_j, \qquad i = 1, 2, \ldots, n. \tag{10.16}$$

Next, we show that (10.15) and (10.16) also constitute a sufficient set of conditions for $\mathbf{x}$ to be an admissible solution. We thus show that if a given $\mathbf{x}$ satisfies the conditions in (10.15) and (10.16) then a set of changes exist in the coefficients and constants which satisfy (10.11) and (10.9).

Now, given a vector $\mathbf{x} = (x_j)$ let us write

$$\gamma_i = \frac{c_i - \sum_{j=1}^{n} a_{ij} x_j}{\sum_{j=1}^{n} \varepsilon_{ij} |x_j| + \varepsilon_i}. \tag{10.17}$$

And, on assuming that (10.15) and (10.16) are satisfied by $\mathbf{x}$, it follows that

$$|\gamma_i| \leq 1. \tag{10.18}$$

Then, taking

$$\delta a_{ij} = \gamma_i \varepsilon_i \operatorname{sign}(x_j), \qquad \delta c_i = -\gamma_i \varepsilon_i, \qquad i, j = 1, 2, \ldots, n \tag{10.19}$$

clearly satisfies (10.9) by (10.18) and further satisfies (10.11).

For, clearly, for each of $i = 1, 2, \ldots, n$,

$$\text{left-hand side of (10.11)} = \sum_{j=1}^{n} \delta a_{ij} x_j - \delta c_i$$

$$= \sum_{j=1}^{n} \gamma_i \varepsilon_{ij} \operatorname{sign}(x_j) x_j + \gamma_i \varepsilon_i, \qquad \text{by (10.19)},$$

$$= \gamma_i \left(\sum_{j=1}^{n} \varepsilon_{ij} |x_j| + \varepsilon_i \right)$$

$$= c_i - \sum_{j=1}^{n} a_{ij} x_j, \qquad \text{by (10.17)},$$

$$= \text{right-hand side of (10.11)}.$$

Hence, (10.15) and (10.16) constitute a sufficient set of conditions for $\mathbf{x}$ to be an admissible solution.

Combining our results, then, it follows that (10.15) and (10.16) constitute a necessary and sufficient set of conditions for $\mathbf{x}$ to be an admissible solution of $\mathbf{A}^*\mathbf{x} = \mathbf{c}^*$ subject to (10.1) and (10.2).

Now let us rewrite (10.15) and (10.16) in the following more convenient form:

$$\sum_{j=1}^{n} (a_{ij} x_j + \varepsilon_{ij} |x_j|) \geq c_i - \varepsilon_i, \qquad i = 1, 2, \ldots, n, \tag{10.20}$$

and

$$\sum_{j=1}^{n} (a_{ij} x_j - \varepsilon_{ij} |x_j|) \leq c_i + \varepsilon_i, \qquad i = 1, 2, \ldots, n. \tag{10.21}$$

And we may regard the necessary and sufficient conditions in (10.20) and (10.21) as specifying the solution set X.

Thus, by (10.20) and (10.21) the solution set of $\mathbf{A}^*\mathbf{x} = \mathbf{c}^*$ subject to (10.1) and (10.2) may be written in place of (10.4) as

$$X = \{\mathbf{x} \mid \mathbf{A}\mathbf{x} + \Delta\mathbf{A}\,|\mathbf{x}| \geq \mathbf{c} - \Delta\mathbf{c},\ \mathbf{A}\mathbf{x} - \Delta\mathbf{A}\,|\mathbf{x}| \leq \mathbf{c} + \Delta\mathbf{c}\}. \quad (10.22)$$

It is clearly easy enough to check whether a given vector $\mathbf{x}$ is an admissible solution. But finding the intervals of uncertainty is another matter.

Now, in Chapter 11 we shall show how to find the intervals of uncertainty by transforming the conditions in (10.20) and (10.21) into standard linear programming form. Determining the intervals of uncertainty would then require carrying out the computational schemes of linear programming.

But in this chapter we are not involved with the standard linear programming form.

To obtain the essence of the rest of this chapter the reader may at this stage note the results in (10.57), (10.67), (10.94), (10.102), and (10.103) and then proceed directly to Section 10.8.

10.3 A Two-dimensional Example

We consider the system of equations

$$x_1 - 2x_2 = -4$$
$$x_1 + 2x_2 = 8 \quad (10.23)$$

with uncertainties

$$\varepsilon_{ij} = \varepsilon_i = 0{\cdot}1 = \varepsilon, \qquad i, j = 1, 2. \quad (10.24)$$

Let $\mathbf{x}^0$ be the solution of

$$\mathbf{A}\mathbf{x}^0 = \mathbf{c} \quad (10.25)$$

where $\mathbf{A}$ is the midpoint coefficient matrix and $\mathbf{c}$ is the midpoint right-hand column of constants (see (10.1)), the notation for the solution now being different from that previously used (see (1.4)). Thus, in our example $\mathbf{x}^0$ is the solution of (10.23) and it is easy to check that

$$x_1^0 = 2, \qquad x_2^0 = 3. \quad (10.26)$$

Since the admissible solution $\mathbf{x}^0$ in (10.26) lies in the first quadrant, let us in the first place determine the subset of solutions in this quadrant. Now, since

$$x_1 \geq 0, \qquad x_2 \geq 0 \quad (10.27)$$

in the first quadrant we may write

$$|x_1| = x_1, \qquad |x_2| = x_2. \tag{10.28}$$

Hence, removing the modulus signs in (10.20) and (10.21) these conditions become in our case

$$\sum_{j=1}^{2} (a_{ij} + \varepsilon_{ij})x_j \geq c_i - \varepsilon_i, \qquad i = 1, 2, \tag{10.29}$$

and

$$\sum_{j=1}^{2} (a_{ij} - \varepsilon_{ij})x_j \leq c_i + \varepsilon_i, \qquad i = 1, 2. \tag{10.30}$$

And by (10.23) and (10.24) we then have

$$
\begin{array}{ll}
1 \cdot 1x_1 - 1 \cdot 9x_2 \geq -4 \cdot 1 & \text{(a)} \\
1 \cdot 1x_1 + 2 \cdot 1x_2 \geq 7 \cdot 9 & \text{(b)}
\end{array}
\tag{10.31}
$$

and

$$
\begin{array}{ll}
0 \cdot 9x_1 - 2 \cdot 1x_2 \leq -3 \cdot 9 & \text{(c)} \\
0 \cdot 9x_1 + 1 \cdot 9x_2 \leq 8 \cdot 1 & \text{(d)}
\end{array}
\tag{10.32}
$$

Thus, (10.31) and (10.32) are the necessary and sufficient conditions for admissible solutions in the first quadrant for our example.

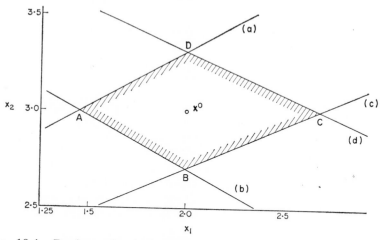

FIG. 10.1   Region of admissible solutions for (10.23) and (10.24) represented by boundary and interior of *ABCD*

Now, each of the conditions in (10.31) and (10.32) divides the first quadrant into two regions. Each condition excludes admissible solutions on the one side of the line representing it in Fig. 10.1 and permits admissible solutions on the other side of the line as well as on the line itself.

Referring, then, to Fig. 10.1, it is clear that all points on the boundary or within the shaded area $ABCD$ satisfy the inequalities in (10.31) and (10.32). And all points in the first quadrant outside the area $ABCD$ do not satisfy all the conditions in (10.31) and (10.32) and hence cannot represent admissible solutions.

Thus, the boundary and interior of the region $ABCD$ represent all the admissible solutions in the first quadrant.

We must now ask ourselves whether there are any admissible solutions in the other quadrants.

Applying the conditions in (10.20) and (10.21) to the other quadrants, it can be shown that they contain no admissible solutions.

Alternatively, one can use the result in (10.104) which implies that the set of all solutions corresponds to one continuous region if the system is not critically ill-conditioned, i.e., the solution set X is then not disjoint.

Now, for our present example the coefficient matrix $\mathbf{A}$ and its inverse $\mathbf{B}$ are

$$\mathbf{A} = \begin{bmatrix} 1 & -2 \\ 1 & 2 \end{bmatrix}, \quad \mathbf{B} = \begin{bmatrix} 0\cdot50 & 0\cdot50 \\ -0\cdot25 & 0\cdot25 \end{bmatrix}. \tag{10.33}$$

Hence for $\varepsilon = 0\cdot1$ given in (10.24) the ill-conditioning factor

$$\phi = \left(\sum_{i=1}^{2} \sum_{j=1}^{2} |b_{ij}| \right)\varepsilon = 0\cdot15 < 1.$$

Thus, by (2.46) the approximate system of equations in (10.23) and (10.24) is not critically ill-conditioned.

Hence the solution set X is not disjoint.

It therefore follows that the boundary and interior of $ABCD$ in Fig. 10.1 represent the entire solution set X.

Still another way of satisfying ourselves that $ABCD$ represents the entire solution set is to use the result in (10.131). In our case, (10.131) merely requires that $\varepsilon_1 \neq 0, \varepsilon_2 \neq 0$, this being clearly satisfied by (10.24).

And it now follows from Fig. 10.1 that the minimum value of x_1 for an admissible solution is the x_1-coordinate of A and that the maximum value of x_1 for an admissible solution is the x_1-coordinate of C. These two values, then, give the endpoints of the interval of uncertainty in x_1. And, similarly, the endpoints of the interval of uncertainty in x_2 are given by the x_2-coordinate of B and the x_2-coordinate of D.

Thus,

$$U_1 = [1\cdot454, 2\cdot667], \qquad U_2 = [2\cdot714, 3\cdot316] \qquad (10.34)$$

(see (10.5)).

Clearly, then, to obtain the intervals of uncertainty above requires determining the coordinates of the points of intersection A, B, C, and D and then taking the minimum and maximum x_1-coordinates and the minimum and maximum x_2-coordinates.

Now, to obtain the coordinates of each of the four points of intersection A, B, C, and D, we can proceed as follows.

We replace the inequality signs in (10.31) and (10.32) by equality signs and then choose every set of two equations from the two sets in (10.31) and (10.32) except that if the ith equation is chosen from the one set then the corresponding equation is not chosen from the other set. (This exception excludes finding the point of intersection of two parallel lines when all the ε_{ij} are zero, with uncertainties existing only in the right-hand constants.)

This then leads to the following four pairs of equations obtained respectively from (a) and (b), (a) and (d), (b) and (c), and (c) and (d) in (10.31) and (10.32):

$$\begin{aligned} 1\cdot1x_1 - 1\cdot9x_2 &= -4\cdot1 \\ 1\cdot1x_1 + 2\cdot1x_2 &= 7\cdot9, \end{aligned} \qquad (10.35)$$

$$\begin{aligned} 1\cdot1x_1 - 1\cdot9x_2 &= -4\cdot1 \\ 0\cdot9x_1 + 1\cdot9x_2 &= 8\cdot1, \end{aligned} \qquad (10.36)$$

$$\begin{aligned} 1\cdot1x_1 + 2\cdot1x_2 &= 7\cdot9 \\ 0\cdot9x_1 - 2\cdot1x_2 &= -3\cdot9, \end{aligned} \qquad (10.37)$$

$$\begin{aligned} 0\cdot9x_1 - 2\cdot1x_2 &= -3\cdot9 \\ 0\cdot9x_1 + 1\cdot9x_2 &= 8\cdot1. \end{aligned} \qquad (10.38)$$

The solutions of the systems of equations (10.35) to (10.38) lead to the coordinates of the four points of intersection (in fact to the coordinates of the vertices A, D, B, and C, respectively, of the polygon).

Now, the solutions of (10.35) to (10.38) are respectively

$$\begin{aligned} x_1 &= 1\cdot454 & x_2 &= 3\cdot000, \\ x_1 &= 2\cdot667 & x_2 &= 3\cdot000, \\ x_1 &= 2\cdot000 & x_2 &= 2\cdot714, \\ x_1 &= 2\cdot000 & x_2 &= 3\cdot316. \end{aligned}$$

Hence,

$$\begin{aligned} U_1 &= [\min(x_1), \max(x_1)] = [1\cdot454, 2\cdot667] \\ U_2 &= [\min(x_2), \max(x_2)] = [2\cdot714, 3\cdot316], \end{aligned} \qquad (10.39)$$

these being the results in (10.34).

And we may mention that in the linear programming approach of the next chapter the intervals in (10.39) can be obtained without any recourse to diagrams.

Solution Set for $\varepsilon = 0\cdot5$

Now, we may expect that the region representing the solution set X increases in size as ε in (10.24) increases.

Thus, had we taken

$$\varepsilon_{ij} = \varepsilon_i = 0\cdot5 = \varepsilon, \qquad i, j = 1, 2 \qquad (10.40)$$

in place of $\varepsilon = 0\cdot1$ in (10.24), we would have obtained the region $ABCD$ in Fig. 10.2.

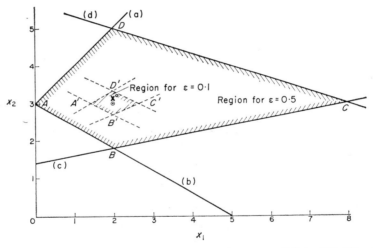

FIG. 10.2   The region $ABCD$ corresponds to $\varepsilon = 0\cdot5$ in (10.40) while the region $A'B'C'D'$ corresponds to $\varepsilon = 0\cdot1$ in (10.24)

We note that one of the points of the region of admissible solutions, A, now lies on the boundary of the quadrant. And with the uncertainties given in (10.40) the intervals of uncertainty for (10.23) are

$$U_1 = [0, 8\cdot00], \qquad U_2 = [1\cdot80, 5\cdot00]. \qquad (10.41)$$

(Compare (10.24) and (10.34) with (10.40) and (10.41).)

Now, if we choose ε to be larger than that in (10.40), it is to be expected that part of the solution set X will lie outside the first quadrant. And we consider such a case in the next section.

10.4 Solution Set in More Than One Quadrant

Before choosing further values of ε for investigation, we first determine the condition for no critical ill-conditioning of the system of equations in (10.23) when the uncertainties in the coefficients are all equal to ε.

Condition for No Critical Ill-conditioning of (10.23)

Now the determinant of the coefficient matrix

$$\mathbf{A} = \begin{bmatrix} 1 & -2 \\ 1 & 2 \end{bmatrix} \tag{10.42}$$

in (10.33) is 4. And it is clear from the distribution of signs in the elements of $\mathbf{A}$ that the determinant can become negative only if some of the elements change sign. Further, it is clear that the determinant can become zero only when some of the elements reach or pass through zero.

Hence, the system is not critically ill-conditioned if

$$\varepsilon < 1, \tag{10.43}$$

for the element of smallest magnitude in $\mathbf{A}$ is 1.

And indeed, if $\varepsilon = 1$ we can find a singular coefficient matrix. For clearly $\det(\mathbf{A}^*) = 0$ where

$$\mathbf{A}^* = \begin{bmatrix} 0 & -2 \\ 0 & 2 \end{bmatrix} \in [\mathbf{A} - \Delta\mathbf{A},\ \mathbf{A} + \Delta\mathbf{A}] \tag{10.44}$$

(see (10.1)).

Hence,

$$\varepsilon < 1 \tag{10.45}$$

is a necessary and sufficient condition for no critical ill-conditioning for the system of equations in (10.23) when the uncertainties in the coefficients are all equal to ε.

But we should note that it is not necessary to determine whether a system is critically ill-conditioned or not before applying the necessary and sufficient conditions for admissible solutions. Indeed, it will be seen that the conditions for admissible solutions can lead to some finite intervals of uncertainty even when the system is critically ill-conditioned.

Nevertheless, knowing that a system of equations is not critically ill-conditioned can be helpful (see (10.104)).

Solution Set for ε = 0·7

In place of (10.24) and (10.40) we now choose

$$\varepsilon_{ij} = \varepsilon_i = 0\cdot7 = \varepsilon, \qquad i = 1, 2. \qquad (10.46)$$

Then firstly we restrict ourselves to the first quadrant, i.e., to the quadrant in which $\mathbf{x}^0$ lies (see (10.26)).

Now, for this quadrant

$$x_1 \geq 0, \qquad x_2 \geq 0, \qquad (10.47)$$

and the conditions corresponding to (10.31) and (10.32) are

$$
\begin{array}{lll}
1\cdot7x_1 - 1\cdot3x_2 \geq -4\cdot7 & \text{(a)} & \\
1\cdot7x_1 + 2\cdot7x_2 \geq 7\cdot3 & \text{(b)} & \\
0\cdot3x_1 - 2\cdot7x_2 \leq -3\cdot3 & \text{(c)} & (10.48) \\
0\cdot3x_1 + 1\cdot3x_2 \leq 8\cdot7 & \text{(d)} &
\end{array}
$$

(see (10.28) to (10.30)).

The conditions in (10.48) clearly lead to the region $ABCD$ in Fig. 10.3, the x_2-axis being intercepted at E and F. But by (10.47) we can accept only that portion of $ABCD$ in the first quadrant as representing admissible solutions.

Thus, $EBCDF$ represents the solution subset in the first quadrant.

But by (10.45) and (10.46) the system of equations is not critically ill-conditioned. Hence, by (10.104) the solution set is not disjoint. But the solution subset in the first quadrant contains solutions for which

$$x_1 = 0, \qquad x_2 > 0 \qquad (10.49)$$

(the line EF). Hence, we have a clear indication that the second quadrant may contain admissible solutions.

Now to find the admissible solutions in the second quadrant, i.e., for

$$x_1 \leq 0, \qquad x_2 \geq 0, \qquad (10.50)$$

we must put

$$|x_1| = -x_1 \qquad \text{and} \qquad |x_2| = x_2 \qquad (10.51)$$

in the necessary and sufficient conditions for admissible solutions in (10.20) and (10.21).

Then, corresponding to (10.29) and (10.30) we now have

$$(a_{i1} - \varepsilon_{i1})x_1 + (a_{i2} + \varepsilon_{i2})x_2 \geq c_i - \varepsilon_i, \qquad i = 1, 2, \quad (10.52)$$

and

$$(a_{i1} + \varepsilon_{i1})x_1 + (a_{i2} - \varepsilon_{i2})x_2 \leq c_i + \varepsilon_i, \qquad i = 1, 2. \quad (10.53)$$

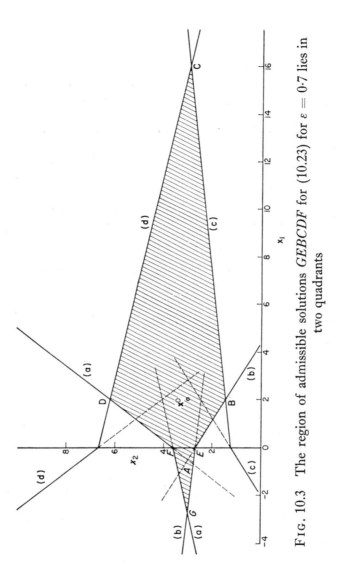

FIG. 10.3 The region of admissible solutions *GEBCDF* for (10.23) for $\varepsilon = 0.7$ lies in two quadrants

This then leads to the following inequalities for (10.23) and (10.46):

$$
\begin{array}{lll}
0 \cdot 3x_1 - 1 \cdot 3x_2 \geq -4 \cdot 7 & \text{(a)} \\
0 \cdot 3x_1 + 2 \cdot 7x_2 \geq 7 \cdot 3 & \text{(b)} \\
1 \cdot 7x_1 - 2 \cdot 7x_2 \leq -3 \cdot 3 & \text{(c)} \\
1 \cdot 7x_1 + 1 \cdot 3x_2 \leq 8 \cdot 7 & \text{(d).}
\end{array}
\qquad (10.54)
$$

Thus, (10.54) are the necessary and sufficient conditions to be satisfied in the second quadrant corresponding to the conditions in (10.48) for the first quadrant.

The conditions in (10.54) subject to (10.50) give GEF in Fig. 10.3 as the region of admissible solutions in the second quadrant. (Note that EF is common to the regions of admissible solutions in the first and second quadrants, the conditions in (10.48) and (10.54) becoming identical for $x_1 = 0$.)

But we have seen that the solution set is not disjoint because the system of equations is not critically ill-conditioned for $\varepsilon = 0 \cdot 7$.

Hence, there are no admissible solutions in the third and fourth quadrants.

Hence, the entire solution set X is represented by the boundary and interior of the region $GEBCDF$. And this conclusion could have been obtained directly by (10.131), for, clearly, by (10.46) $\varepsilon_1 = \varepsilon_2 = 0 \cdot 7 \neq 0$.

The intervals of uncertainty are now obtained from the coordinates of

$$
\begin{array}{ll}
G = (-2 \cdot 67, \, 3 \cdot 00), & B = (2 \cdot 00, \, 1 \cdot 44), \\
C = (16 \cdot 00, \, 3 \cdot 00), & D = (2 \cdot 00, \, 6 \cdot 33),
\end{array}
\qquad (10.55)
$$

the intervals of uncertainty being

$$
U_1 = [-2 \cdot 67, \, 16 \cdot 00], \qquad U_2 = [1 \cdot 44, \, 6 \cdot 33]. \qquad (10.56)
$$

Next, we note a feature concerning the nature of solution sets of approximate systems of equations which Fig. 10.3 illustrates, namely that:

> Solution sets of approximate systems of equations are not necessarily convex and can be nonconvex. $\qquad (10.57)$

For, by definition, a set is convex if for every two elements $\mathbf{x}_1$ and $\mathbf{x}_2$ in the set

$$
\mu \mathbf{x}_1 + (1 - \mu) \mathbf{x}_2 \qquad (10.58)
$$

is also an element in the set where

$$
0 \leq \mu \leq 1. \qquad (10.59)
$$

And in the case of two and three dimensions it can be shown that this requires that all points on the straight line joining $\mathbf{x}_1$ to $\mathbf{x}_2$ are also in the set.

Now, in Fig. 10.3 it is clear that no point between G and D on the straight line joining G to D is an element of the solution set. Hence (10.57) holds.

On the other hand, the solution sets corresponding to Figs 10.1 and 10.2 are convex.

We next show that in the above example (i.e. (10.23) with $\varepsilon = 0.7$) the only available approach of finding the intervals of uncertainty is in fact that of using the conditions for admissible solutions.

Methods I to VIII for $\varepsilon = 0.7$

Let us first consider those methods that lead to intervals containing the true intervals of uncertainty.

Now, for

$$\mathbf{B} = \begin{bmatrix} 0.50 & 0.50 \\ -0.25 & 0.25 \end{bmatrix}, \quad \mathbf{\Delta A} = \begin{bmatrix} 0.7 & 0.7 \\ 0.7 & 0.7 \end{bmatrix}, \tag{10.60}$$

the ill-conditioning matrix

$$\mathbf{C} = |\mathbf{B}|\, \mathbf{\Delta A} = \begin{bmatrix} 0.70 & 0.70 \\ 0.35 & 0.35 \end{bmatrix} \tag{10.61}$$

(see (10.33) and (1.37)).

Hence,

$$\phi_{\text{norm}} \equiv \|\mathbf{C}\|_{\text{I}} = 1.40 > 1 \tag{10.62}$$

(see (3.15) and (2.38)).

Therefore Method I is inapplicable because its condition of application is $\phi_{\text{norm}} < 1$. And in fact the condition of application of Method I is only satisfied for the system in (10.23) provided $\varepsilon < 0.50$.

Now, the ill-conditioning factor

$$\phi = \left(\sum_{i=1}^{2} \sum_{j=1}^{2} |b_{ij}| \right)\varepsilon = 1.05 > 1 \tag{10.63}$$

(see (1.36), (2.46), and (6.50)).

Hence, Method IV cannot be applied. And we may note that to satisfy the condition $\phi < 1$ in Method IV requires that $\varepsilon < 2/3$.

And Method VII, the interval arithmetic method in which the coefficient matrix has been strongly diagonalized, is found to fail at $\varepsilon = 2/3$, it being noted that this is the same value of ε for which Method IV fails. And ordinary method interval arithmetic (O.M.) is found to fail as expected at an even smaller value of ε. In fact, O.M. fails at $\varepsilon = 0.5$.

Further, Methods VI and VIII cannot be used for $\varepsilon = 0.7$ because they require the use of Methods IV and VII.

We are therefore left with Methods II, III, and V.

Now, Method II should not be used unless the ill-conditioning factor is considerably less than unity. And if Method III is to lead to meaningful results it should obviously not be used when Methods I, IV, and VII fail. Finally, Method V is inapplicable because (7.29) requires that $\phi < \frac{1}{2}$. And, in any case, (7.5) implies that Method V cannot be applied when Method IV is inapplicable.

Thus, the only approach available to us when $\varepsilon = 0.7$ is that of using the necessary and sufficient conditions for admissible solutions in (10.20) and (10.21). And it is in this way that we found the intervals of uncertainty in (10.56).

Next, we show that our present approach even leads to useful information when the system of equations is critically ill-conditioned (see (10.45)).

Solution Set for $\varepsilon = 1$

We now determine the solution set for the system of equations in (10.23) as it becomes critically ill-conditioned, i.e., for

$$\varepsilon_{ij} = \varepsilon_i = 1 = \varepsilon, \qquad i, j = 1, 2 \qquad (10.64)$$

(see (10.45)).

Now, subject to (10.64) the conditions (10.20) and (10.21) for admissible solutions in the first quadrant become

$$\begin{array}{rll} 2x_1 - x_2 \geq -5 & \text{(a)} & \\ 2x_1 + 3x_2 \geq 7 & \text{(b)} & \\ -3x_2 \leq -3 & \text{(c)} & (10.65) \\ x_2 \leq 9 & \text{(d).} & \end{array}$$

And for $x_1 \leq 0$, $x_2 \geq 0$, i.e., in the second quadrant, the conditions to be satisfied by admissible solutions are

$$\begin{array}{rll} -x_2 \geq -5 & \text{(a)} & \\ 3x_2 \geq 7 & \text{(b)} & \\ 2x_1 - 3x_2 \leq -3 & \text{(c)} & (10.66) \\ 2x_1 + x_2 \leq 9 & \text{(d).} & \end{array}$$

Now, the region of admissible solutions in the first two quadrants satisfying (10.65) and (10.66) is shown in Fig. 10.4. And we may check by (10.20), (10.21), and (10.64) that the solution set does not extend into the third and fourth quadrants. Hence, the solution set lies only in the first two quadrants.

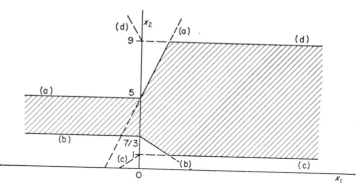

FIG. 10.4   The solution set for (10.23) for $\varepsilon = 1$. The set is unbounded in x_1 in both directions

This example clearly shows that:

> When a system becomes critically ill-conditioned then the intervals of uncertainty for some of the unknowns can remain bounded although the interval of uncertainty for at least one unknown becomes unbounded, for otherwise the system would not be critically ill-conditioned.   (10.67)

For in the example, while the interval of uncertainty for x_1 is unbounded that for x_2 is clearly bounded, the interval for x_2 being

$$U_2 = [1, 9].\qquad(10.68)$$

And here we note that of the previous methods which lead to intervals containing the true intervals of uncertainty, none would give finite uncertainties in any of the unknowns if the system is critically ill-conditioned.

Thus, procedures based directly on the necessary and sufficient conditions for admissible solutions have clear advantages in extreme cases.

10.5 Solution Set of Example in (1.43) and (1.44)

So far, we have failed to obtain the true interval of uncertainty in x_2 by any of the previous methods for the example in (1.43) and (1.44), i.e., for

$$0.53x_1 + 0.86x_2 + 0.48x_3 = 0.64$$
$$0.94x_1 - 0.47x_2 + 0.85x_3 = 3.12 \qquad (10.69)$$
$$0.87x_1 + 0.55x_2 + 0.26x_3 = 0.83$$

with the uncertainties

$$\varepsilon_{ij} = \varepsilon_i = 0.01 = \varepsilon, \qquad i, j = 1, 2, 3. \qquad (10.70)$$

We now determine the exact intervals of uncertainty in the above approximate system of equations by applying the conditions for admissible solutions in (10.20) and (10.21).

Now, denoting the solution of (10.69) by $\mathbf{x}^0$ ($\mathbf{Ax}^0 = \mathbf{c}$) we find that to 12D

$$x_1^0 = 0.972996913984, \qquad x_2^0 = -0.996151894376, \\ x_3^0 = 2.043754718232. \qquad (10.71)$$

Let us, then, determine the extent of the solution set in the octant for which

$$x_1 \geq 0, \qquad x_2 \leq 0, \qquad x_3 \geq 0, \qquad (10.72)$$

i.e., in the octant containing $\mathbf{x}^0$.

The conditions in (10.20) and (10.21) then become

$$0.54x_1 + 0.85x_2 + 0.49x_3 \geq 0.63 \qquad \text{(a)}$$
$$0.95x_1 - 0.48x_2 + 0.86x_3 \geq 3.11 \qquad \text{(b)} \qquad (10.73)$$
$$0.88x_1 + 0.54x_2 + 0.27x_3 \geq 0.82 \qquad \text{(c)}$$

and

$$0.52x_1 + 0.87x_2 + 0.47x_3 \leq 0.65 \qquad \text{(a}')$$
$$0.93x_1 - 0.46x_2 + 0.84x_3 \leq 3.13 \qquad \text{(b}') \qquad (10.74)$$
$$0.86x_1 + 0.56x_2 + 0.25x_3 \leq 0.84 \qquad \text{(c}').$$

Now, each of the conditions in (10.73) and (10.74) is equivalent to a plane dividing the octant into regions where admissible solutions may and may not lie. The resultant region for admissible solutions in the orthant in (10.72) thus consists of the boundary and interior of the region formed by the six dividing planes in (10.73) and (10.74) and by the three dividing planes in (10.72), it being noted that the n-dimensional term *orthant* corresponds to the two-dimensional term *quadrant* and the three-dimensional *octant* and that it includes both these terms. But let us assume for our example that no admissible solution in fact lies on the boundary planes in (10.72) and that the resultant region for admissible solutions in the orthant defined by (10.72) consists of the boundary and interior of the six-sided parallelepiped formed by the six dividing planes in (10.73) and (10.74). By (10.131) and (10.70) this would then represent the entire solution set.

To determine the intervals of uncertainty in the unknowns therefore requires determining the coordinates of the eight corners (vertices) of the parallelepiped, each such point being the point of intersection of three planes. And if no such coordinate is zero and if no such coordinate lies outside the orthant in (10.72), then our assumption in the previous paragraph that no admissible solution lies on the boundary planes of the orthant is valid.

We clearly have to solve a system of three linear algebraic equations to determine the coordinates of a vertex.

Adopting a procedure similar to that in Section 10.3 (see (10.35) to (10.38)), we now replace the inequality signs in (10.73) and (10.74) by equality signs. And then we write down every set of three equations that can be chosen from (10.73) and (10.74) except that if we include the ith equations from (10.73) we do not include the corresponding equation from (10.74).

This, then, leads in all to the following eight sets (2^3 sets) of equations, and these will give us the coordinates of the eight vertices of the parallelepiped:

Set 1, corresponding to (a), (b), (c) in (10.73) and (10.74)

$$
\begin{aligned}
0 \cdot 54 x_1 + 0 \cdot 85 x_2 + 0 \cdot 49 x_3 &= 0 \cdot 63 \\
0 \cdot 95 x_1 - 0 \cdot 48 x_2 + 0 \cdot 86 x_3 &= 3 \cdot 11 \\
0 \cdot 88 x_1 + 0 \cdot 54 x_2 + 0 \cdot 27 x_3 &= 0 \cdot 82,
\end{aligned}
\tag{10.75}
$$

Set 2, corresponding to (a), (b), (c') in (10.73) and (10.74)

$$
\begin{aligned}
0 \cdot 54 x_1 + 0 \cdot 85 x_2 + 0 \cdot 49 x_3 &= 0 \cdot 63 \\
0 \cdot 95 x_1 - 0 \cdot 48 x_2 + 0 \cdot 86 x_3 &= 3 \cdot 11 \\
0 \cdot 86 x_1 + 0 \cdot 56 x_2 + 0 \cdot 25 x_3 &= 0 \cdot 84,
\end{aligned}
\tag{10.76}
$$

Set 3, corresponding to (a), (b'), (c') in (10.73) and (10.74)

$$
\begin{aligned}
0 \cdot 54 x_1 + 0 \cdot 85 x_2 + 0 \cdot 49 x_3 &= 0 \cdot 63 \\
0 \cdot 93 x_1 - 0 \cdot 46 x_2 + 0 \cdot 84 x_3 &= 3 \cdot 13 \\
0 \cdot 86 x_1 + 0 \cdot 56 x_2 + 0 \cdot 25 x_3 &= 0 \cdot 84,
\end{aligned}
\tag{10.77}
$$

Set 4, corresponding to (a), (b'), (c) in (10.73) and (10.74)

$$
\begin{aligned}
0 \cdot 54 x_1 + 0 \cdot 85 x_2 + 0 \cdot 49 x_3 &= 0 \cdot 63 \\
0 \cdot 93 x_1 - 0 \cdot 46 x_2 + 0 \cdot 84 x_3 &= 3 \cdot 13 \\
0 \cdot 88 x_1 + 0 \cdot 54 x_2 + 0 \cdot 27 x_3 &= 0 \cdot 82,
\end{aligned}
\tag{10.78}
$$

Set 5, corresponding to (a'), (b'), (c') in (10.73) and (10.74)

$$0\cdot52x_1 + 0\cdot87x_2 + 0\cdot47x_3 = 0\cdot65$$
$$0\cdot93x_1 - 0\cdot46x_2 + 0\cdot84x_3 = 3\cdot13 \qquad (10.79)$$
$$0\cdot86x_1 + 0\cdot56x_2 + 0\cdot25x_3 = 0\cdot84,$$

Set 6, corresponding to (a'), (b'), (c) in (10.73) and (10.74)

$$0\cdot52x_1 + 0\cdot87x_2 + 0\cdot47x_3 = 0\cdot65$$
$$0\cdot93x_1 - 0\cdot46x_2 + 0\cdot84x_3 = 3\cdot13 \qquad (10.80)$$
$$0\cdot88x_1 + 0\cdot54x_2 + 0\cdot27x_3 = 0\cdot82,$$

Set 7, corresponding to (a'), (b), (c) in (10.73) and (10.74)

$$0\cdot52x_1 + 0\cdot87x_2 + 0\cdot47x_3 = 0\cdot65$$
$$0\cdot95x_1 - 0\cdot48x_2 + 0\cdot86x_3 = 3\cdot11 \qquad (10.81)$$
$$0\cdot88x_1 + 0\cdot54x_2 + 0\cdot27x_3 = 0\cdot82,$$

Set 8, corresponding to (a'), (b), (c') in (10.73) and (10.74)

$$0\cdot52x_1 + 0\cdot87x_2 + 0\cdot47x_3 = 0\cdot65$$
$$0\cdot95x_1 - 0\cdot48x_2 + 0\cdot86x_3 = 3\cdot11 \qquad (10.82)$$
$$0\cdot86x_1 + 0\cdot56x_2 + 0\cdot25x_3 = 0\cdot84.$$

The solutions in Table 10.1 of the systems of equations (10.75) to (10.82) then give the coordinates of the eight vertices of the parallelepiped.

TABLE 10.1 Coordinates of Vertices of Parallelepiped for (1.43) and (1.44)†

System of equations	Solutions		
	x_1	x_2	x_3
Set 1 (10·75)	0·936721027785	−1·015389781483	2·014799916850
Set 2 (10·76)	1·106636358969	−1·015196336167	1·827210718161
Set 3 (10·77)	1·117789691172	−1·066800768201	1·904437183139
Set 4 (10·78)	0·943091852228	−1·067191299883	2·097638785097
Set 5 (10·79)	1·009950897082	−0·976554397815	2·073250765140
Set 6 (10·80)	0·835863544790	−0·976609546923	2·265960133047
Set 7 (10·81)	0·832472054825	−0·927585469745	2·178965723765
Set 8 (10·82)	1·001809696664	−0·927707195304	1·991838760956

† The table gives the coordinates of the vertices of the parallelepiped whose boundary and interior give the solution set of (1.43) and (1.44) (or (10.69) and (10.70)). The minimum and maximum values in each column are underlined.

Now, the solution set of an approximate system of equations is not disjoint for a non-critically ill-conditioned system of equations (see (10.104)). But $\phi = 0 \cdot 08611 < 1$ by (1.48) (see (2.46)). Hence, since all the x_i in Table 10.1 lie within the octant being considered and none on its boundary (see (10.72)) it follows that the entire solution set of (1.43) and (1.44) is represented by the boundary and interior of the parallelepiped whose coordinates are given in the table. Alternatively, we can come to the same conclusion more directly by (10.131) and (10.70).

It now follows from Table 10.1 that the intervals of uncertainty in the unknowns of the approximate system of equations in (1.43) and (1.44) are

$$
\begin{aligned}
U_1 &= [(x_1)_{\min}, (x_1)_{\max}] = [0 \cdot 832472054825, 1 \cdot 117789691172] \\
U_2 &= [(x_2)_{\min}, (x_2)_{\max}] = [-1 \cdot 067191299883, -0 \cdot 927585469745] \\
U_3 &= [(x_3)_{\min}, (x_3)_{\max}] = [1 \cdot 827210718161, 2 \cdot 265960133047].
\end{aligned}
$$
$$(10.83)$$

And the widths of the intervals of uncertainty in (10.83) are

$$
\begin{aligned}
w(U_1) &= 0 \cdot 285317636347 \\
w(U_2) &= 0 \cdot 139605830134 \\
w(U_3) &= 0 \cdot 438749414886.
\end{aligned}
$$
$$(10.84)$$

First, we may now check that the solution in (10.71) of the midpoint system of equations in (1.43) and (1.44) is included in the intervals in (10.83).

And for record purposes we now give the uncertainties in the unknowns as defined in (1.11) and (1.14).

We have by (1.11) and (10.5) with a slight change in notation that

$$
x_i^0 - e_i = (x_i)_{\min}, \qquad x_i^0 + d_i = (x_i)_{\max}, \qquad i = 1, 2, \ldots, n,
$$

so that

$$
e_i = x_i^0 - (x_i)_{\min}, \qquad d_i = (x_i)_{\max} - x_i^0, \qquad i = 1, 2, \ldots, n. \tag{10.85}
$$

Hence, since by (1.14) $\Delta x_i = \max(e_i, d_i)$ we have by (10.71), (10.83), and (10.85) that

$$
\begin{aligned}
\Delta x_1 &= \max(0 \cdot 140524859159, 0 \cdot 144792777188) = 0 \cdot 144792777188 \\
\Delta x_2 &= \max(0 \cdot 071039405507, 0 \cdot 068566424631) = 0 \cdot 071039405507 \\
\Delta x_3 &= \max(0 \cdot 216544000071, 0 \cdot 222205414815) = 0 \cdot 222205414815.
\end{aligned}
$$
$$(10.86)$$

Second, we may check that the systems of equations which lead to columns 1, 2, and 3 of **D** and to columns 1, 2, and 3 of **E** in (4.69) or

(5.73) are those in (10.77), (10.82), (10.80), (10.81), (10.78), and (10.76), respectively.

Hence, let us note that the maximum value of x_2 is in fact the value of x_2 in the first column of E and not of the diagonal element in the second column of D, as would have been the case had the assumptions in (5.1) been valid.

Thus, the conditions for admissible solutions lead to the true intervals of uncertainty to any required precision in all the unknowns of our example, and in particular in x_2. All the previous methods failed to lead to the true interval of uncertainty in x_2.

10.6 Comparison of Results of Methods I to VIII

Using the true intervals of uncertainty in (10.83) for the example in (1.43) and (1.44), we now compare the results by Methods I to VIII.

TABLE 10.2   Comparison of Intervals of Uncertainty by Methods
I to VIII†

Method	endpoints for x_1		endpoints for x_2		endpoints for x_3	
	lower	upper	lower	upper	lower	upper
I	0·766332	0·751374	2·063908	2·081622	0·325081	0·312177
II	0·007367	−0·007591	−0·009466	0·008248	0·006368	−0·006535
III	0	0	0	−0·000872	0	0
IV	0·054471	0·039513	0·037591	0·055305	0·053475	0·040572
V	−0·039737	−0·054695	−0·056523	−0·038809	−0·040738	−0·053642
VI	0	0	0·000050	0·000016	0	0
VII	0·043266	0·039513	0·037591	0·046587	0·005672	0·040572
VIII	0	0	0·000050	0·000016	0	0

† The table gives the differences between the endpoints of the intervals of uncertainty by the various methods and the true endpoints in (10.83). A positive sign indicates that an endpoint obtained by the method being considered lies outside the true interval of uncertainty and a negative sign that it lies inside. The distance that an endpoint lies from the corresponding true endpoint is given as a fraction of the length of the true interval of uncertainty.

The results are given in Table 10.2 and the footnote to the table explains its interpretation.

By way of further explanation, we show how the entries for x_1 by Method II are obtained.

By (4.35) and (10.71) the interval of uncertainty by Method II for x_1 is

$$U_1^{II} = [0 \cdot 9730 - 0 \cdot 1426, 0 \cdot 9730 + 0 \cdot 1426] = [0 \cdot 8304, 1 \cdot 1156].$$
(10.87)

But by (10.83) and (10.84) the true interval of uncertainty for x_1 is

$$U_1 = [0 \cdot 8325, 1 \cdot 1178]$$
(10.88)

and the width is

$$w(U_1) = 0 \cdot 2853.$$
(10.89)

Hence, by (10.87) and (10.88) the lower endpoint in (10.87) is outside the true interval of uncertainty at a distance

$$0 \cdot 8325 - 0 \cdot 8304 = 0 \cdot 0021$$
(10.90)

from the lower endpoint of the true interval of uncertainty. The positive value in (10.90) thus indicates that the lower endpoint of U_1^{II} lies outside the true interval of uncertainty U_1.

Finally, we express the value in (10.90) as a fraction of the length of the true interval of uncertainty in (10.89)

$$0 \cdot 0021 / 0 \cdot 2853 = 0 \cdot 0074,$$
(10.91)

this being the entry in Table 10.2 for the lower endpoint of x_1 by Method II.

And the entry in Table 10.2 for the upper endpoint is obtained from (10.87), (10.88), and (10.89), as follows

$$(1 \cdot 1156 - 1 \cdot 1178) / 0 \cdot 2853 = -0 \cdot 0022 / 0 \cdot 2853 = -0 \cdot 0077 \quad (10.92)$$

(actually, $-0 \cdot 007591$ in the table), the negative sign indicating that the upper endpoint as computed by Method II is inside the true interval of uncertainty.

Referring to Table 10.2, then, it is clear that the results by Methods IV and VII do not differ by much. And both give errors of under ten per cent in the widths of the intervals for our example. But the intervals by Method VII are slightly shorter than those by Method IV.

The improved versions of Method IV and VII, namely Methods VI and VIII, seem to be equally good. The improvement in precision is considerable, the error in the width of the interval of uncertainty in x_2 being $0 \cdot 0066$ per cent, the true intervals of uncertainty being obtained for x_1 and x_3. And the interval arithmetic Method VIII gives marginally better results (see (9.115)).

One does not want to generalize from an example, but perhaps it may be said that seldom would better results be required than those by Methods VIII and VI, especially as these methods give bounds on the endpoints of the intervals of uncertainty (see (9.114) and (8.65), for example).

However, we have seen that the true intervals of uncertainty can be obtained by satisfying the conditions for admissible solutions in (10.20) and (10.21). And we shall show in the next chapter how the problem can be put into standard linear programming form.

10.7   Interdependence of Components in Solution Vector

Let us recall that we may regard the conditions for admissible solutions in (10.20) and (10.21) as defining the solution set X (see (10.22)). And we have interested ourselves in determining the minimum and maximum values of each component of $\mathbf{x}$ in X:

$$U_i = [(x_i)_{\min}, (x_i)_{\max}], \qquad i = 1, 2, \ldots, n \qquad (10.93)$$

(see (10.5)).

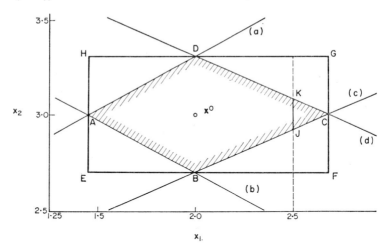

FIG. 10.5   This figure is a variation of Fig. 10.1. Note that the rectangle $EFGH$ defined by the intervals of uncertainty contains the region of uncertainty $ABCD$. And note that if x_1 is specified as 2·5, say, then x_2 can only lie in the interval represented by JK within the interval of uncertainty in x_2

Referring now to Fig. 10.5, it is clear that for the example in (10.23) and (10.24) the intervals of uncertainty determine the rectangle $EFGH$

containing the solution set X represented by $ABCD$ (see Fig. 10.1 and also (10.34)). But we note that not all points in the rectangle $EFGH$ represent admissible solutions; only those points in or on the boundary of $ABCD$ represent admissible solutions.

It is thus clear that an admissible solution is not necessarily one whose components fall within the intervals of uncertainty. It is certainly a necessary condition for the components of an admissible solution to fall within the intervals of uncertainty but not a sufficient condition.

Thus:

> For a vector **x** to be an admissible solution it is necessary but not sufficient that all its components fall within the (10.94) intervals of uncertainty.

Nevertheless, suppose we are free to specify the values of one or more of the unknowns as we might be able to do in design problems. Then, as far as approximate equation theory is concerned, we can specify any one unknown of our choice to lie anywhere within its interval of uncertainty (or within any interval within its interval of uncertainty).

But having specified one component of a solution vector, this restricts in general the intervals in which the others may lie, these restricted intervals being of course contained in the original intervals of uncertainty. And as further components of the solution vector are specified, so the intervals in which the remaining unknowns may lie become further restricted.

This, then, explains the title of this section.

For example, for (10.23) and (10.24) the intervals of uncertainty are given in (10.34):

$$U_1 = [1{\cdot}454,\, 2{\cdot}667], \qquad U_2 = [2{\cdot}714,\, 3{\cdot}316] \qquad (10.95)$$

(see Figs 10.1 and 10.5).

But suppose we can specify the value of x_1 as $2{\cdot}500$. Then the interval in which x_2 may now lie is represented in Fig. 10.5 by JK, the intersection of the region of uncertainty $ABCD$ and the line $x_1 = 2{\cdot}500$. And this interval is

$$\bar{U}_2 = [2{\cdot}929,\, 3{\cdot}079], \qquad \text{so that} \qquad w(\bar{U}_2) = 0{\cdot}150, \qquad (10.96)$$

the width of the new 'interval of uncertainty' $\bar{U}_2$ being much smaller than the width of the interval of uncertainty in x_2 in (10.95) which is

$$w(U_2) = 3{\cdot}316 - 2{\cdot}714 = 0{\cdot}602. \qquad (10.97)$$

Let us now note that the new interval in (10.96) in which x_2 may lie would in general be obtained from the conditions for admissible solutions in (10.20) and (10.21) and not graphically.

Thus the value of any one unknown that can be specified would be substituted in the conditions for admissible solutions. Or, in another situation, the values of any unknowns that become known would be substituted in the conditions for admissible solutions. Then we could obtain the intervals in which the remaining unknowns may lie from the resulting inequalities.

For example, substituting

$$x_1 = 2 \cdot 5 \qquad (10.98)$$

in (10.31) and (10.32) we obtain

$$
\begin{aligned}
2 \cdot 75 - 1 \cdot 9x_2 &\geq -4 \cdot 1 \\
2 \cdot 75 + 2 \cdot 1x_2 &\geq 7 \cdot 9 \\
2 \cdot 25 - 2 \cdot 1x_2 &\leq -3 \cdot 9 \\
2 \cdot 25 + 1 \cdot 9x_2 &\leq 8 \cdot 1.
\end{aligned}
\qquad (10.99)
$$

This, then, leads to

$$
\begin{aligned}
-1 \cdot 9x_2 &\geq -6 \cdot 85 \\
2 \cdot 1x_2 &\geq 5 \cdot 15 \\
-2 \cdot 1x_2 &\leq -6 \cdot 15 \\
1 \cdot 9x_2 &\leq 5 \cdot 85,
\end{aligned}
$$

i.e., to

$$x_2 \leq 3 \cdot 605, \quad x_2 \geq 2 \cdot 452, \quad x_2 \geq 2 \cdot 929, \quad x_2 \leq 3 \cdot 079. \quad (10.100)$$

Clearly the interval in x_2 satisfying (10.100) is

$$\bar{U}_2 = [2 \cdot 929, 3 \cdot 079], \qquad (10.101)$$

the conditions $x_2 \leq 3 \cdot 605$ and $x_2 \geq 2 \cdot 452$ in (10.100) being redundant.

We have thus obtained both graphically and analytically the new interval in which x_2 may lie if $x_1 = 2 \cdot 5$, the general approach being of course the analytical one (see (10.96) and (10.101)).

Summarizing, then:

> Any one unknown of one's own choice may be specified to lie anywhere in its interval of uncertainty as far as approximate equation theory is concerned. But the intervals in which the remaining unknowns lie may then become restricted. These intervals can be obtained by satisfying the conditions for admissible solutions afresh with the value of the specified unknown substituted in the conditions.   (10.102)

And in conclusion we state the basic result which gives rise to this section:

> Admissible solutions may lie only in the region of uncertainty corresponding to the solution set X but not in the region which is not common to the region of uncertainty and the region defined by the intervals of uncertainty. Or, put in another way, the solution set X is a subset of the set defined by the intervals of uncertainty.   (10.103)

Now, in the next three sections we derive certain properties concerning the continuity and boundedness of the solution set and we give one further condition on critical ill-conditioning. It will be seen that this new condition can succeed where our previous sufficient conditions on critical ill-conditioning may fail. Further, it will be seen that this new condition is particularly well suited when the intervals of uncertainty are obtained by using the conditions for admissible solutions.

10.8   Continuity of Solution Set

In this section, we prove an important property concerning the continuity of the solution set X:

> The solution set X of an approximate system of equations is not disjoint if the system is not critically ill-conditioned.   (10.104)

And we have seen in Sections 10.3 and 10.4 that the above result enables one in certain situations to conclude that a subset which has been determined is in fact the entire solution set.

To prove (10.104) we first prove a property that we have tacitly assumed throughout:

> The components of the solution vector $\mathbf{x}$ of a system of equations $\mathbf{Ax} = \mathbf{c}$ are continuous functions of the elements of $\mathbf{A}$ and $\mathbf{c}$, provided that $\det(\mathbf{A})$ does not become zero, or, briefly, the solution vector $\mathbf{x}$ of a system of equations $\mathbf{Ax} = \mathbf{c}$ is a continuous function of $\mathbf{A}$ and $\mathbf{c}$ provided $\det(\mathbf{A})$ does not become zero   (10.105)

(see (4.1), for example).

Now it is clear that:

> The determinant of any square matrix $\mathbf{A}$ is a continuous function of the elements of $\mathbf{A}$.   (10.106)

A L A E—M

For the determinant is equal to the sum of a number of terms each of which is a product of a number of elements of **A**.

But by Cramer's rule the kth component of the solution vector **x** of **Ax** = **c** is given by

$$x_k = \frac{\det(\mathbf{A}_k)}{\det(\mathbf{A})}, \qquad (10.107)$$

where $\mathbf{A}_k$ is a matrix in which the kth column of **A** is replaced by the column of constants **c**.

It now follows by (10.106) and (10.107) that x_k is a continuous function of the elements of **A** and **c**, provided that $\det(\mathbf{A})$ in the denominator of (10.107) does not become zero.

The result in (10.105) follows.

And the result in (10.104) then follows.

For let $\mathbf{x}_1$ and $\mathbf{x}_2$ be any two solution vectors of X. These vectors must then satisfy (10.20) and (10.21). Hence, it is possible to find two matrices $\mathbf{A}_1$ and $\mathbf{A}_2$ and two vectors $\mathbf{c}_1$ and $\mathbf{c}_2$, such that

$$\mathbf{A}_1\mathbf{x}_1 = \mathbf{c}_1, \qquad \mathbf{A}_2\mathbf{x}_2 = \mathbf{c}_2 \qquad (10.108)$$

where

$$\mathbf{A}_1, \mathbf{A}_2 \in \mathbf{A}^{\mathrm{I}}, \qquad \mathbf{c}_1, \mathbf{c}_2 \in \mathbf{c}^{\mathrm{I}} \qquad (10.109)$$

(see (10.1), (10.17), and (10.19)).

Now, let us proceed from $\mathbf{A}_1$ and $\mathbf{c}_1$ along any continuous path in the $(n + 1)n$-dimensional space of **A** and **c** to $\mathbf{A}_2$ and $\mathbf{c}_2$, the path being restricted to lie in the interval of uncertainty $\mathbf{A}^{\mathrm{I}}$ in the coefficients and the interval of uncertainty $\mathbf{c}^{\mathrm{I}}$ in the right-hand constants (see (10.1)). It then follows by (10.105) that the solution vector traces out some continuous path joining $\mathbf{x}_1$ to $\mathbf{x}_2$ if we assume that the system of equations is not critically ill-conditioned, i.e., if we assume that

$$\det(\mathbf{A}) \neq 0 \qquad \text{for all } \mathbf{A} \in \mathbf{A}^{\mathrm{I}}. \qquad (10.110)$$

The result in (10.104) thus clearly follows.

But let us note that if a system of equations is critically ill-conditioned then the solution set X may be disjoint.

For example, consider the one-dimensional equation

$$[-3, 5]x = 30. \qquad (10.111)$$

This equation is critically ill-conditioned because

$$0 \in [-3, 5]. \qquad (10.112)$$

And the solution set X of (10.112) is clearly disjoint, the solution set being the union of the two subsets X_1 and X_2 where

$$X_1 = \{x \mid x \leq -10\} \quad \text{and} \quad X_2 = \{x \mid x \geq 6\}. \quad (10.113)$$

Now in concluding this section let us note the following result that follows from (10.105):

> The inverse $\mathbf{B}$ of a matrix $\mathbf{A}$ is a continuous function of $\mathbf{A}$, provided det($\mathbf{A}$) does not become zero. $\qquad (10.114)$

For it is well known that the kth column of $\mathbf{B}$, $k = 1, 2, \ldots, n$, is the solution vector of $\mathbf{Ax} = \mathbf{c}$, when $\mathbf{c}$ is the kth unit vector, i.e., the vector whose kth component is unity and all other components zero (see Table 1.3, for example).

Having dealt with some continuity properties of the solution set, we next deal with the boundedness of the solution set.

10.9 Boundedness of Solution Set

In this section we first prove the following result which we have tacitly assumed to hold:

> The solution set of a non-critically ill-conditioned system of equations is bounded. $\qquad (10.115$

And it then follows that:

> If a solution set is unbounded in any one direction in solution space (i.e., in any one component of the solution vector), then the system is critically ill-conditioned. $\qquad (10.116)$

And then we prove the following result which involves both continuity and boundedness:

> Commencing at any admissible solution of a critically ill-conditioned system of equations, it is always possible to trace out a continuous path in solution space which is unbounded in at least one direction in this space, provided the uncertainties in all the right-hand constants are nonzero, i.e., provided $\varepsilon_i \neq 0$, $i = 1, 2, \ldots, n$. $\qquad (10.117)$

To prove (10.115), let us then consider a non-critically ill-conditioned system of equations. Then we have by definition of a non-critically ill-conditioned system that

$$\det(\mathbf{A}) \neq 0 \qquad \text{for} \quad \mathbf{A} \in \mathbf{A}^{\mathrm{I}} \tag{10.118}$$

where $\mathbf{A}^{\mathrm{I}}$ is the interval coefficient matrix.

Further, $\det(\mathbf{A})$ does not approach zero because $\mathbf{A}^{\mathrm{I}}$ contains all its limit points, $\mathbf{A}^{\mathrm{I}}$ being taken to be closed by definition in (10.1). Hence, for a non-critically ill-conditioned system of equations

$$\det(\mathbf{A}) \neq 0 \qquad \text{and} \qquad \det(\mathbf{A}) \nrightarrow 0 \qquad \text{for} \quad \mathbf{A} \in \mathbf{A}^{\mathrm{I}}. \tag{10.119}$$

Thus, if a system is not critically ill-conditioned

$$|\det(\mathbf{A})| \geq \delta > 0 \tag{10.120}$$

for some δ, the modulus sign being introduced because $\det(\mathbf{A})$ may be negative. (It follows by (10.106) and (10.118) that $\det(\mathbf{A})$ is always of the same sign for all $\mathbf{A} \in \mathbf{A}^{\mathrm{I}}$ for a given non-critically ill-conditioned system of equations.)

Now, let us note that $\det(\mathbf{A}_k)$ in (10.107) is bounded because the elements of $\mathbf{A}$ and $\mathbf{c}$ are bounded by their finite intervals of uncertainty. Hence, by (10.107) and (10.120) it follows that the component x_k is bounded. And this holds for $k = 1, 2, \ldots, n$.

We have thus proved (10.115).

And (10.116) then clearly follows.

Next we prove (10.117).

We now consider a critically ill-conditioned system of equations and we restrict ourselves to the case where each right-hand constant has a nonzero uncertainty, i.e.,

$$\Delta \mathbf{c} = (\varepsilon_i), \qquad \varepsilon_i \neq 0, \qquad i = 1, 2, \ldots, n. \tag{10.121}$$

Now, let $\mathbf{x}_1$ be any solution in X. Then $\mathbf{x}_1$ satisfies (10.20) and (10.21). Hence, it is possible to find a matrix $\mathbf{A}_1$ and a vector $\mathbf{c}_1$ such that

$$\mathbf{A}_1 \mathbf{x}_1 = \mathbf{c}_1 \qquad \text{for some} \qquad \mathbf{A}_1 \in \mathbf{A}^{\mathrm{I}} \quad \text{and} \quad \mathbf{c}_1 \in \mathbf{c}^{\mathrm{I}} \tag{10.122}$$

(see (10.1)).

Let us now proceed from $\mathbf{A}_1$ and $\mathbf{c}_1$ along a continuous path in the $(n+1)n$-dimensional space of $\mathbf{A}^{\mathrm{I}}$ and $\mathbf{c}^{\mathrm{I}}$ until some matrix $\mathbf{A}_2$ is approached for which $\det(\mathbf{A}_2) = 0$. Since we are considering a critically ill-conditioned system there is at least one $\mathbf{A} \in \mathbf{A}^{\mathrm{I}}$ for which $\det(\mathbf{A}) = 0$.

Then we can show that at least one component of $\mathbf{x}$ becomes unbounded as $\mathbf{A} \rightarrow \mathbf{A}_2$.

For as $\mathbf{A}$ approaches $\mathbf{A_2}$

$$\det(\mathbf{A}) \to 0. \tag{10.123}$$

Hence, writing $\mathbf{B} = \mathbf{A}^{-1}$ we then have that

$$\det(\mathbf{B}) \to \infty \tag{10.124}$$

because

$$\det(\mathbf{A}) \det(\mathbf{B}) = 1 \tag{10.125}$$

if $\mathbf{B} = \mathbf{A}^{-1}$.

But $\det(\mathbf{B})$ is equal to the sum of a number of terms each of which is the product of a number of elements of $\mathbf{B}$. Hence:

> As $\det(\mathbf{A}) \to 0$ so at least one element of its inverse $\mathbf{B}$ becomes unbounded. (10.126)

Hence, writing $\mathbf{Ax} = \mathbf{c}$ where $\mathbf{A} \in \mathbf{A}^I$, $\mathbf{c} \in \mathbf{c}^I$, as

$$\mathbf{x} = \mathbf{Bc}, \tag{10.127}$$

it follows that:

> At least one element of $\mathbf{x}$ becomes unbounded as $\det(\mathbf{A}) \to 0$ for some $\mathbf{c} \in \mathbf{c}^I$, subject to $\varepsilon_i \neq 0$, $i = 1, 2, \ldots, n$ (see (10.121)). (10.128)

For, reverting to our previous notation for the moment where $\mathbf{c}$ is the midpoint column of right-hand constants (see (10.1)), we can write (10.127) as $\mathbf{x} = \mathbf{B}(\mathbf{c} + \delta\mathbf{c}) = (\mathbf{Bc} + \mathbf{B}\delta\mathbf{c})$ where $|\delta\mathbf{c}| \leq \Delta\mathbf{c}$. Now, if $\mathbf{Bc}$ in $(\mathbf{Bc} + \mathbf{B}\delta\mathbf{c})$ becomes unbounded then we need merely choose $\delta\mathbf{c} = \mathbf{0}$. But if $\mathbf{Bc}$ in $(\mathbf{Bc} + \mathbf{B}\delta\mathbf{c})$ remains bounded let us suppose that the element b_{km} of the inverse becomes unbounded as $\mathbf{A_2}$ is approached (see (10.126)). Then the component x_k in the solution vector becomes unbounded as $\mathbf{A_2}$ is approached if we choose $\delta c_m = \varepsilon_m$, $\delta c_i = 0$, $i \neq m$ (see (10.121)). Hence (10.128) holds.

The proof of (10.117) then clearly follows by (10.105) and (10.128). But let us note that:

> The condition $\varepsilon_i \neq 0$, $i = 1, 2, \ldots, n$, in (10.117) is a sufficient condition but not a necessary one. (10.129)

For in the proof we required only one component of $\delta\mathbf{c}$ to be nonzero. Also, if the element b_{km} of the inverse becomes unbounded and the column of right-hand constants is the mth unit vector, then x_k becomes unbounded with $\delta\mathbf{c} = \mathbf{0}$. Clearly, (10.129) holds.

However, the condition $\varepsilon_i \neq 0$, $i = 1, 2, \ldots, n$ in (10.129) is not in general a redundant one. For example, if $\mathbf{c}$ in $(\mathbf{c} + \delta\mathbf{c})$ is the zero vector, then (10.128) and (10.117) would not hold if $\Delta\mathbf{c} = \mathbf{0}$. For then $\delta\mathbf{c} = \mathbf{0}$, so that both $\mathbf{Bc}$ and $\mathbf{B}\delta\mathbf{c}$ would be zero in

$$\mathbf{x} = \mathbf{B}(\mathbf{c} + \delta\mathbf{c}) = \mathbf{Bc} + \mathbf{B}\delta\mathbf{c},$$

irrespective of the behaviour of the elements of $\mathbf{B}$.

Next, we use the results in (10.117) and (10.129) to obtain further sufficient conditions on no critical ill-conditioning.

10.10   Conditions for No Critical Ill-conditioning

First, by way of terminology, let us say that a subset of the solution set is *insular* if it is a closed bounded subset containing at least one admissible solution and surrounded by a region in which there are no admissible solutions. For example, the subset $ABCD$ in Fig. 10.1 is insular as is the subset $GEBCDF$ in Fig. 10.3. But, on the other hand, we cannot say that the subset $ABCD$ in Fig. 10.2 is insular until we have investigated the solution set in the second quadrant. We would then in fact find that the only admissible solution in the second quadrant is at A. We could then say that $ABCD$ is an insular subset of the solution set, for we are now satisfied that $ABCD$ is surrounded by a region not containing admissible solutions.

More precisely, then, an *insular subset* X_1 of the solution set X is a nonempty closed bounded subset of the solution set which can be contained in an open set of the space such that the complement of X_1 and the open set contain no admissible solutions.

Using this terminology then we shall prove that:

> If an insular subset X_1 of the solution set X exists, then the system is not critically ill-conditioned subject to $\varepsilon_i \neq 0$, $i = 1, 2, \ldots, n$. (10.130)

And it then follows by (10.104) that:

> If an insular subset X_1 of the solution set X exists then X_1 is the entire solution set, i.e., $X_1 = X$, subject to $\varepsilon_i \neq 0$, $i = 1, 2, \ldots, n$. (10.131)

Now, to prove (10.130) we need merely say that subject to $\varepsilon_i \neq 0$, $i = 1, 2, \ldots, n$, an insular subset X_1 of the solution set X cannot exist by (10.117) for a critically ill-conditioned system of equations. And (10.131) then follows by (10.104).

And it is also clear by (10.117) that:

> If the solution set of an approximate system of equations is bounded then the system is not critically ill-conditioned, subject to $\varepsilon_i \neq 0$, $i = 1, 2, \ldots, n$. (10.132)

For, clearly, the assumption that the midpoint coefficient matrix $\mathbf{A}$ is nonsingular (see Section 1.1 and equations (1.4) and (1.5)) is equivalent to:

> In our treatment an approximate system of equations is assumed to have at least one admissible solution, namely, $\mathbf{x}^0$ (see (10.25)). (10.133)

Thus (10.132) follows from (10.117).
But let us note that:

> The condition $\varepsilon_i \neq 0$, $i = 1, 2, \ldots, n$, in (10.130), (10.131), and (10.132) is a sufficient condition but not a necessary one. (10.134)

For this clearly follows by (10.129) and the use of (10.117) in proving (10.130), (10.131), and (10.132).

Now, the results in (10.130) and (10.134) and in (10.131) and (10.134) may be looked at in different ways.

On the one hand, we can regard (10.130) and (10.134) as a sufficient condition for no critical ill-conditioning. And it can be more powerful than the sufficient conditions

$$\rho(\mathbf{C}) < 1, \qquad \|\mathbf{C}\|_{\mathrm{I}} < 1, \qquad \|\mathbf{C}\|_{\mathrm{II}} < 1, \qquad (10.135)$$

in the sense that it can indicate no critical ill-conditioning where the conditions in (10.135) fail to do so (see (2.36)).

For example, the conditions in (10.135) fail for the system of equations in (10.23) for $\varepsilon_{ij} = \varepsilon_i = 0 \cdot 7 = \varepsilon$, $i, j = 1, 2$ (see (10.62), (10.63), and (2.46)). But, since it is clear from Fig. 10.3 that an insular subset of the solution set exists, it follows by (10.130) that the system is not critically ill-conditioned, the condition $\varepsilon_1 \neq 0$, $\varepsilon_2 \neq 0$ being satisfied (see (10.46)).

Thus (10.130) may be effective when the other sufficient conditions for no critical ill-conditioning fail. And at the same time we can obtain the intervals of uncertainty from the insular subset X_1 (see Fig. 10.3 and equation (10.56)).

On the other hand, we may say that (10.131) and (10.134) imply that if the conditions for admissible solutions lead to an insular subset subject to the sufficient (but not necessary) condition

$$\varepsilon_i \neq 0, \qquad i = 1, 2, \ldots, n, \qquad (10.136)$$

then this is the entire set and there is no need to concern ourselves with the question of critical ill-conditioning.

Hence in particular we have that:

> If, on restricting our investigation of the conditions
> for admissible solutions to an orthant containing at
> least one admissible solution, we obtain finite 'inter-
> vals of uncertainty' with no zero endpoints, then, sub-
> ject to the sufficient condition in (10.136), the
> intervals are the true intervals of uncertainty.   (10.137)

For then there clearly exists an insular subset of the solution set in the orthant concerned.

10.11 Conclusions

We have seen the clear advantages in extreme cases of finding the intervals of uncertainty directly from the conditions for admissible solutions. Thus, in some situations Methods I to VIII are all not applicable, but the solution set can nevertheless be obtained from the conditions for admissible solutions (Section 10.4). And the conditions for admissible solutions can lead to useful information even when the system is critically ill-conditioned (see (10.67)).

But we have seen that Methods VI and VIII may be expected in most situations to give sufficiently satisfactory estimates of the intervals of uncertainty (Section 10.6). And an important feature of these two methods is that they provide bounds on the endpoints of the intervals of uncertainty. And they lead to the true intervals of uncertainty when they reduce to Method III (see (8.11) and (9.100)).

It is the conditions for admissible solutions, however, that always lead to the true intervals of uncertainty. And in the next chapter we show how the problem of obtaining the intervals of uncertainty from the conditions for admissible solutions can be put into standard linear programming form.

CHAPTER 11

Method IX: Intervals of Uncertainty by Linear Programming

11.1 Introduction

We have seen in Chapter 10 that finding the intervals of uncertainty from the conditions for admissible solutions has the clear advantage that it always leads to the true intervals of uncertainty. And the computational scheme to be used in practice is one of the procedures in linear programming. Method IX then involves obtaining the true intervals of uncertainty from the conditions for admissible solutions by using one of the computational schemes of linear programming (l.p.).

And if one of the unknowns is specified then l.p. can be used to find the restricted intervals in which the remaining unknowns lie (see Section 10.7). Thus, l.p. can also be used to obtain a clearer picture of the solution set in certain circumstances than that given by the intervals of uncertainty alone.

But linear programming is clearly a more specialized field of knowledge than that of solving linear algebraic equations by Gaussian elimination, even when this is carried out in interval arithmetic. Nevertheless, a standard linear programming program (or subroutine) could be used which would not require any specialized knowledge about the computational aspects of l.p.

In Section 11.2 we deal with the standard formulation of a problem to be solved by an l.p. subroutine. And in Sections 11.3 to 11.7 we deal with reducing our problem to standard l.p. form, giving numerical examples. And in Sections 11.8 to 11.10 we consider the situation when the solution set extends over more than one orthant.

Thus, our approach is to reduce our problem to standard l.p. form but not to concern ourselves with the construction of an l.p. subroutine.

Some l.p. subroutine may be expected to be included as standard software in a computing centre.

11.2 The Standard Formulation of the l.p. Problem

The general linear programming (l.p.) problem is to find a vector

$$\mathbf{x} = (x_1, x_2, \ldots, x_q)' \tag{11.1}$$

which minimizes the linear form, the so-called *objective function*,

$$c_1 x_1 + c_2 x_2 + \ldots + c_q x_q \tag{11.2}$$

subject to the *nonnegativity conditions*

$$x_i \geq 0 \qquad i = 1, 2, \ldots, q \tag{11.3}$$

and to the *constraints*

$$a_{11} x_1 + a_{12} x_2 + \ldots + a_{1q} x_q = b_1$$
$$a_{21} x_1 + a_{22} x_2 + \ldots + a_{2q} x_q = b_2 \tag{11.4}$$
$$\ldots$$
$$a_{p1} x_1 + a_{p2} x_2 + \ldots + a_{pq} x_q = b_p$$

where the a_{ij}, b_i, and c_i are given constants.

And it is usual to assume that the equations in (11.4) have been multiplied by -1 where necessary so as to make all the

$$b_i \geq 0, \qquad i = 1, 2, \ldots, p. \tag{11.5}$$

It is clearly a trivial matter to satisfy the conditions in (11.5) and we shall assume that these are always satisfied.

Thus (11.1) to (11.5) is the standard formulation of an l.p. problem. And using computer programming language our approach is that if we enter the values of the a_{ij}, the b_i, and the c_i then the l.p. subroutine will return the minimum value of the objective function. And we may reasonably assume that if required the values of the x_i making the objective function a minimum are available.

Now, for brevity let us write the standard formulation of the l.p. problem in matrix notation.

Let us write

$$\mathbf{c} = (c_1, c_2, \ldots, c_q)$$
$$\mathbf{A} = (a_{ij}) \qquad i = 1, 2, \ldots, p, \quad j = 1, 2, \ldots, q,$$
$$\mathbf{b} = (b_1, b_2, \ldots, b_p)'.$$

Then we can write the objective function (11.2) as

$$\mathbf{cx}, \tag{11.6}$$

the scalar product of $\mathbf{c}$ and $\mathbf{x}$. And we can write the constraints (11.4) as

$$\mathbf{Ax} = \mathbf{b}. \tag{11.7}$$

Thus:

> The l.p. problem in standard form is to find a vector $\mathbf{x}$ which minimizes the objective function $\mathbf{cx}$ subject to the nonnegativity conditions $\mathbf{x} \geq \mathbf{0}$ and the constraints $\mathbf{Ax} = \mathbf{b}$, it being assumed that $\mathbf{b} \geq \mathbf{0}$ (see paragraph below (11.5)). (11.8)

The Minimum and Maximum Values of x_k

We now show how we can obtain the minimum and maximum values of x_k, the kth component of $\mathbf{x}$, subject to the nonnegativity conditions $\mathbf{x} \geq \mathbf{0}$ and the constraints $\mathbf{Ax} = \mathbf{b}$.

Suppose we take

$$c_k = 1, \qquad c_i = 0, \quad i \neq k, \tag{11.9}$$

so that the objective function (11.2) merely becomes

$$x_k. \tag{11.10}$$

Then, since solving the l.p. problem leads to the minimum value of the objective function, we in fact obtain the minimum value of x_k. And this minimum value can be obtained by the l.p. subroutine returning either the value of the objective function or the value of the component x_k.

Next, suppose we take

$$c_k = -1, \qquad c_i = 0, \quad i \neq k, \tag{11.11}$$

so that our objective function becomes

$$-x_k. \tag{11.12}$$

Then, solving the l.p. problem leads to the minimum value of the objective function, i.e., to the minimum value of $-x_k$. Hence minus the value of the objective function is equal to the maximum value of x_k. Thus the value of x_k in this case is equal to the maximum value of x_k subject to (11.3) and (11.4).

Summarizing, then, we have:

> With the objective function defined in turn by (11.9) and (11.11), the values obtained for x_k are, respectively, the minimum and maximum values of x_k, subject to the nonnegativity conditions $\mathbf{x} \geq \mathbf{0}$ and the constraints $\mathbf{Ax} = \mathbf{b}$. (11.13)

It is thus clear that:

> To find the minimum and maximum values of each of n, say, of the q components of $\mathbf{x} = (x_1, x_2, \ldots, x_q)'$ subject to (11.3) and (11.4) requires calling the l.p. subroutine $2n$ times. $\qquad$ (11.14)

We next consider the problem of obtaining the intervals of uncertainty in the unknowns of an approximate system of equations by reducing our requirements to standard linear programming form.

11.3   Reducing Our Problem to Standard l.p. Form

There are two aspects that require attention:

1. We have inequalities in the conditions for admissible solutions in (10.20) and (10.21) rather than equalities as in (11.4).

2. Our interest lies not only in the orthant for which all the $x_i \geq 0$ as in (11.3). In fact, we are frequently interested only in the orthant in which $\mathbf{x}^0$ lies (see (10.25)).

Once the above two aspects have been attended to, it will be seen that:

> If in investigating the solution set X in the orthant containing $\mathbf{x}^0$ (see (10.25)) it becomes evident that the entire solution set lies in this orthant then we have in fact obtained the n intervals of uncertainty by solving $2n$ linear programming problems. $\qquad$ (11.15)

(See (11.14) and (10.131).)

And we now deal with the two aspects mentioned above in the next two sections.

11.4   Introduction of Slack Variables

A condition involving an inequality can be changed to one involving an equality by introducing a so-called *slack variable*.

For example, the condition or constraint

$$3x_1 + 4x_2 \leq 5 \qquad (11.16)$$

may be written as

$$3x_1 + 4x_2 + x_3 = 5 \qquad (11.17)$$

where

$$x_3 \geq 0. \qquad (11.18)$$

And the variable x_3 introduced to change the inequality in (11.16) to an equality in (11.17) is called a *slack variable*.

We now note by (11.18) that the slack variable x_3 satisfies the non-negativity conditions in (11.3).

But suppose that the inequality involves a $\geq$ sign rather than a $\leq$ as in (11.16). Then the slack variable must be introduced with opposite sign to that in (11.17) for the nonnegativity conditions to be satisfied.

For example, suppose we require to put the constraint

$$3x_1 + 4x_2 \geq 5 \tag{11.19}$$

into standard linear programming form. Then we would write

$$3x_1 + 4x_2 - x_3 = 5 \tag{11.20}$$

where

$$x_3 \geq 0. \tag{11.21}$$

It is thus always possible to remove inequality signs in order to put constraints into standard l.p. form by introducing slack variables. And this can be done in such a way that the nonnegativity conditions are satisfied.

Example

Suppose we wish to determine the intervals of uncertainty for the example in (10.23) and (10.24) by expressing the conditions for admissible solutions in standard l.p. form, the intervals of uncertainty being obtained by calling the l.p. subroutine a number of times.

Then we begin by determining the part of the solution set in the orthant in which $\mathbf{x}^0$ lies, i.e., in the orthant for which

$$x_1 \geq 0, \qquad x_2 \geq 0 \tag{11.22}$$

(see (10.26) and (10.27)).

At this stage then our conditions in (11.22) are compatible with the nonnegativity conditions in (11.3) so that there is no difficulty in this connection.

Now, the conditions for admissible solutions of (10.23) and (10.24), subject to (11.22) (see (10.27)), are given in (10.31) and (10.32). And when the inequality signs are changed to equality signs by introducing slack variables the constraints in (10.31) and (10.32) become

$$
\begin{array}{llll}
1{\cdot}1x_1 - 1{\cdot}9x_2 - x_3 & & = -4{\cdot}1 & \text{(a)} \\
1{\cdot}1x_1 + 2{\cdot}1x_2 & - x_4 & = 7{\cdot}9 & \text{(b)} \\
0{\cdot}9x_1 - 2{\cdot}1x_2 & + x_5 & = -3{\cdot}9 & \text{(c)} \\
0{\cdot}9x_1 + 1{\cdot}9x_2 & + x_6 = & 8{\cdot}1 & \text{(d)}
\end{array} \tag{11.23}
$$

where

$$x_i \geq 0, \qquad i = 3, 4, 5, 6. \tag{11.24}$$

And combining (11.22) and (11.24) we have

$$x_i \geq 0, \qquad i = 1, 2, \ldots, 6. \tag{11.25}$$

Now, multiplying equations (a) and (c) in (11.23) throughout by -1, so as to satisfy (11.5), the constraints in (11.23) become

$$
\begin{array}{llll}
-1 \cdot 1 x_1 + 1 \cdot 9 x_2 + x_3 & = 4 \cdot 1 & \text{(a)} \\
1 \cdot 1 x_1 + 2 \cdot 1 x_2 \quad\quad - x_4 & = 7 \cdot 9 & \text{(b)} \\
-0 \cdot 9 x_1 + 2 \cdot 1 x_2 \quad\quad\quad - x_5 & = 3 \cdot 9 & \text{(c)} \\
0 \cdot 9 x_1 + 1 \cdot 9 x_2 \quad\quad\quad\quad + x_6 & = 8 \cdot 1 & \text{(d).}
\end{array}
\tag{11.26}
$$

Thus corresponding to the general l.p. conditions in (11.3) and (11.4) we have for our example the conditions in (11.25) and (11.26) for admissible solutions in the $\mathbf{x}^0$-orthant.

Now, to determine the minimum and maximum values of x_1 and x_2 subject to (11.25) and (11.26), we define the following four objective functions:

$$
\begin{array}{ll}
c_1 = 1, & c_2 = 0; \\
c_1 = -1, & c_2 = 0; \\
c_1 = 0, & c_2 = 1; \\
c_1 = 0, & c_2 = -1
\end{array}
\tag{11.27}
$$

(see (11.13)).

Then, calling the l.p. subroutine four times we obtain in turn

$$
\begin{array}{l}
x_1 = 1 \cdot 454 = (x_1)_{\min} \\
x_1 = 2 \cdot 667 = (x_1)_{\max} \\
x_2 = 2 \cdot 714 = (x_2)_{\min} \\
x_2 = 3 \cdot 316 = (x_2)_{\max}
\end{array}
\tag{11.28}
$$

(see (10.34)).

But it is well known from l.p. theory that:

The set of all solutions to an l.p. problem is a closed convex set (see (10.58)). (11.29)

(The set is closed because we have $\leq$ and $\geq$ signs in our conditions rather than $<$ and $>$ signs.)

Hence, since all the values in (11.28) are finite and none are zero it follows that an insular subset X_1 of the solution set X exists in the $\mathbf{x}^0$-orthant.

It follows by (10.131) and (10.24) that this insular subset is in fact the entire solution set, i.e.,

$$X_1 = X. \tag{11.30}$$

Hence we have by (11.28) that the intervals of uncertainty of the approximate system of equations in (10.23) and (10.24) are

$$U_1 = [(x_1)_{min}, (x_1)_{max}] = [1\cdot454, 2\cdot667]$$
$$U_2 = [(x_2)_{min}, (x_2)_{max}] = [2\cdot714, 3\cdot316] \tag{11.31}$$

(see (10.34)).

But let us mention that had one of the values in (11.28) been zero, then in view of our restriction $x_i \geq 0$ in (11.25) we would have an indication that the solution set may extend into other orthants. We could then not say that an insular subset exists without appropriate investigation. And in general the solution set would consist of the union of the convex subsets in the various orthants.

The above did not, however, arise in our example.

We next deal with the other aspect mentioned in Section 11.3, namely, the case where we wish to investigate the solution set in an orthant other than the one for which $\mathbf{x} \geq \mathbf{0}$, i.e., other than, let us say, the *positive orthant* (see (11.8)).

11.5   Admissible Solutions in Any Orthant

We show by way of an example how to investigate the solution set by l.p. in an orthant other than the positive one. And then we draw some general conclusions from our example.

Let us then investigate by l.p. the solution set of the example in (1.43) and (1.44) in its $\mathbf{x}^0$-orthant.

We have by (1.47) that the solution of the midpoint system of equations in (1.43) is

$$x_1^0 = 0\cdot9730, \qquad x_2^0 = -0\cdot9962, \qquad x_3^0 = 2\cdot0438. \tag{11.32}$$

Investigating the solution set in the $\mathbf{x}^0$-orthant implies that we are restricting ourselves to

$$x_1 \geq 0, \qquad x_2 \leq 0, \qquad x_3 \geq 0. \tag{11.33}$$

Now, to satisfy the nonnegativity conditions in (11.3) we must first introduce the following change of variable:

$$x_1' = x_1, \qquad x_2' = -x_2, \qquad x_3' = x_3. \tag{11.34}$$

Then

$$x_i' \geq 0, \qquad i = 1, 2, 3. \tag{11.35}$$

And let us note that subject to (11.33) and (11.34)

$$|x_1| = x_1 = x_1', \qquad |x_2| = -x_2 = x_2', \qquad |x_3| = x_3 = x_3', \tag{11.36}$$

these relations being required to remove the modulus signs in (10.20) and (10.21). In this way the otherwise nonlinear conditions for admissible solutions are linearized.

Expressing, then, the conditions for admissible solutions in (10.20) and (10.21) for our example in (1.43) and (1.44) in terms of the new variables in (11.34), we have by (11.36) that

$$
\begin{aligned}
0\cdot54x_1' - 0\cdot85x_2' + 0\cdot49x_3' &\geq 0\cdot63 \\
0\cdot95x_1' + 0\cdot48x_2' + 0\cdot86x_3' &\geq 3\cdot11 \\
0\cdot88x_1' - 0\cdot54x_2' + 0\cdot27x_3' &\geq 0\cdot82 \\
0\cdot52x_1' - 0\cdot87x_2' + 0\cdot47x_3' &\leq 0\cdot65 \\
0\cdot93x_1' + 0\cdot46x_2' + 0\cdot84x_3' &\leq 3\cdot13 \\
0\cdot86x_1' - 0\cdot56x_2' + 0\cdot25x_3' &\leq 0\cdot84.
\end{aligned}
\tag{11.37}
$$

And we note that the constraints in (11.37) are the same as those in (10.73) and (10.74) except that they differ in sign in the corresponding second terms and of course except that the variables are primed (x_i') in (11.37).

Now, converting the inequalities in (11.37) to equalities by introducing slack variables we obtain

$$
\begin{aligned}
0\cdot54x_1' - 0\cdot85x_2' + 0\cdot49x_3' - x_4' \qquad\qquad\qquad &= 0\cdot63 \\
0\cdot95x_1' + 0\cdot48x_2' + 0\cdot86x_3' \qquad - x_5' \qquad\qquad &= 3\cdot11 \\
0\cdot88x_1' - 0\cdot54x_2' + 0\cdot27x_3' \qquad\qquad - x_6' \qquad\quad &= 0\cdot82 \\
0\cdot52x_1' - 0\cdot87x_2' + 0\cdot47x_3' \qquad\qquad\qquad + x_7' \qquad &= 0\cdot65 \\
0\cdot93x_1' + 0\cdot46x_2' + 0\cdot84x_3' \qquad\qquad\qquad\quad + x_8' \quad &= 3\cdot13 \\
0\cdot86x_1' - 0\cdot56x_2' + 0\cdot25x_3' \qquad\qquad\qquad\qquad + x_9' &= 0\cdot84
\end{aligned}
\tag{11.38}
$$

where

$$x_1' \geq 0, \qquad i = 4, 5, \ldots, 9. \tag{11.39}$$

And combining (11.35) with (11.39) we have

$$x_1' \geq 0, \qquad i = 1, 2, \ldots, 9. \tag{11.40}$$

Thus, (11.38) and (11.40) are the linear constraints and nonnegativity conditions for our l.p. problems.

And to obtain the minimum and maximum values of x_1', x_2', and x_3' we

now have to call the l.p. subroutine with the objective function defined
in turn by

$$
\begin{aligned}
c_1 &= 1, & c_2 &= 0, & c_3 &= 0;\\
c_1 &= -1, & c_2 &= 0, & c_3 &= 0;\\
c_1 &= 0, & c_2 &= 1, & c_3 &= 0;\\
c_1 &= 0, & c_2 &= -1, & c_3 &= 0;\\
c_1 &= 0, & c_2 &= 0, & c_3 &= 1;\\
c_1 &= 0, & c_2 &= 0, & c_3 &= -1.
\end{aligned}
\tag{11.41}
$$

The six callings of the l.p. subroutine then lead to

$$
\begin{aligned}
x_1' &= 0{\cdot}8325 = (x_1')_{\min}\\
x_1' &= 1{\cdot}1178 = (x_1')_{\max}\\
x_2' &= 0{\cdot}9276 = (x_2')_{\min}\\
x_2' &= 1{\cdot}0672 = (x_2')_{\max}\\
x_3' &= 1{\cdot}8272 = (x_3')_{\min}\\
x_3' &= 2{\cdot}2660 = (x_3')_{\max}.
\end{aligned}
\tag{11.42}
$$

It now follows by (11.34) that

$$
\begin{aligned}
(x_1)_{\min} &= (x_1')_{\min} = 0{\cdot}8325, & (x_1)_{\max} &= (x_1')_{\max} = 1{\cdot}1178\\
(x_2)_{\min} &= -(x_2')_{\max} = -1{\cdot}0672, & (x_2)_{\max} &= -(x_2')_{\min} = -0{\cdot}9276\\
(x_3)_{\min} &= (x_3')_{\min} = 1{\cdot}8272, & (x_3)_{\max} &= (x_3')_{\max} = 2{\cdot}2660.
\end{aligned}
\tag{11.43}
$$

Since all these values are finite and none are zero, it follows that an
insular subset of the solution set exists in the orthant concerned. It then
follows by (10.131) and (1.44) that this insular subset is the entire
solution set. Hence, by (11.43)

$$
\begin{aligned}
U_1 &= [0{\cdot}8325, \, 1{\cdot}1178]\\
U_2 &= [-1{\cdot}0672, \, -0{\cdot}9276]\\
U_3 &= [1{\cdot}8272, \, 2{\cdot}2660].
\end{aligned}
\tag{11.44}
$$

These, then, are the intervals of uncertainty for our example.

Let us at this stage generalize from our example.

To investigate the solution set in an orthant we in general introduce
a change of variable (see (11.34), for example), no change of variable
being required, however, if the orthant concerned is the positive one
(see example in Section 11.4). And we note that the orthant to be considered first is usually the one known to contain at least one admissible
solution, namely, the orthant containing $\mathbf{x}^0$ (see (10.25)).

The $2n$ conditions for admissible solutions in (10.20) and (10.21) are
then expressed in terms of the new variables and linearized, i.e., without
modulus signs appearing (see (11.36) and (11.37), for example). These,
then, are the $2n$ linear constraints for the orthant in question. Then $2n$

ALAE—N

slack variables are introduced to convert the inequality signs to equality signs (see (11.38), for example).

It is thus clear that:

> In transforming the conditions for admissible solutions of an approximate system of equations of order n to standard l.p. form we have that p and q in (11.4) are given by $p = 2n$, $q = n + 2n = 3n$.
> $\qquad$ (11.45)

And:

> To find the minimum and maximum values of all the components of $\mathbf{x}$ in an orthant requires calling the l.p. subroutine $2n$ times.
> $\qquad$ (11.46)

In the next section we deal more formally with the reduction of the approximate equation problem to l.p. form, introducing the concept of an orthant vector $\mathbf{s}$.

11.6 Orthant Intervals of Uncertainty by Linear Programming

The Orthant Vector

Given any vector $\mathbf{x}$ with nonzero components, let us specify the orthant in which it lies by defining an *orthant vector* $\mathbf{s} = (s_i)$ as follows:

$$s_i = \text{sign}(x_i), \qquad x_i \neq 0, \qquad i = 1, 2, \ldots, n. \qquad (11.47)$$

Thus the components of $\mathbf{s}$ are either $+1$ or -1, it being recalled that the sign function is defined as follows:

$$\text{sign}(a) = +1, -1, 0 \qquad \text{according as } a > 0, a < 0, a = 0. \qquad (11.48)$$

In the definition of an orthant vector, we have avoided the ambiguity that would arise if a component of $\mathbf{x}$ is zero, for the vector $\mathbf{x}$ can then be regarded as lying in more than one orthant. For example, the vector $\mathbf{x} = (0, 3)$ lies both in the first and second quadrants.

To extend the definition we now specify:

> If $x_i = 0$, then take first $s_i = 1$ and then $s_i = -1$, thereby doubling the number of orthants specified.
> $\qquad$ (11.49)

And it clearly follows that:

> A vector $\mathbf{x}$ with m zero components will by (11.47) and (11.49) specify 2^m different orthant vectors, and such a vector $\mathbf{x}$ can be regarded as lying in 2^m different orthants.
> $\qquad$ (11.50)

We may emphasize that having specified an orthant by $\mathbf{s} = (s_i)$ where for any particular i, $s_i = +1$ or $s_i = -1$, the orthant consists of the totality of vectors $\mathbf{x} = (x_i)$ for which

$$|x_i| = s_i x_i, \qquad i = 1, 2, \ldots, n. \tag{11.51}$$

And let us note that all points on the boundaries of the orthant (i.e., vectors with one or more zero components satisfying (11.51)) are therefore regarded as belonging to the orthant.

Reduction to Standard l.p. Form

We now show how to investigate the solution set in an orthant specified by $\mathbf{s} = (s_i)$ where for any particular value of i

$$s_i = +1 \qquad \text{or} \qquad s_i = -1. \tag{11.52}$$

First we introduce the change of variable

$$x_i' = s_i x_i, \qquad i = 1, 2, \ldots, n, \tag{11.53}$$

so that by (11.51) and (11.53)

$$|x_i| = x_i', \qquad i = 1, 2, \ldots, n. \tag{11.54}$$

Thus

$$x_i' = |x_i| \geq 0, \qquad i = 1, 2, \ldots, n. \tag{11.55}$$

And by (11.53) and (11.52) we have that $s_i^2 x_i = s_i x_i'$, i.e.,

$$x_i = s_i x_i', \qquad i = 1, 2, \ldots, n. \tag{11.56}$$

Hence, with the change of variable in (11.53) the conditions for admissible solutions in (10.20) and (10.21) become by (11.54) and (11.56)

$$\sum_{j=1}^{n} (a_{ij} s_j x_j' + \varepsilon_{ij} x_j') \geq c_i - \varepsilon_i, \qquad i = 1, 2, \ldots, n, \tag{11.57}$$

and

$$\sum_{j=1}^{n} (a_{ij} s_j x_j' - \varepsilon_{ij} x_j') \leq c_i - \varepsilon_i, \qquad i = 1, 2, \ldots, n \tag{11.58}$$

(compare with (11.37)).

Converting the inequality signs to equality signs by introducing slack variables we then have

$$\sum_{j=1}^{n} (a_{ij} s_j + \varepsilon_{ij}) x_j' - x_{i+n}' = c_i - \varepsilon_i, \qquad i = 1, 2, \ldots, n,$$

and   (11.59)

$$\sum_{j=1}^{n}(a_{ij}s_j - \varepsilon_{ij})x' + x'_{i+2n} = c_i + \varepsilon_i, \qquad i = 1, 2, \ldots, n,$$

where

$$x'_i \geq 0, \qquad i = n+1, n+2, \ldots, 3n \qquad (11.60)$$

(compare with (11.38) and (11.39)).

And combining (11.55) and (11.60) we have

$$x'_i \geq 0, \qquad i = 1, 2, \ldots, 3n. \qquad (11.61)$$

Now, to obtain the minimum and maximum values of x'_k, say, we take the following objective functions

$$\begin{aligned} c_k &= +1, & c_i &= 0, & i &\neq k; \\ c_k &= -1, & c_i &= 0, & i &\neq k \end{aligned} \qquad (11.62)$$

(see (11.13)). And let us denote the values of x'_k for the above two objective functions by u_k and v_k, respectively, i.e.,

$$u'_k = (x'_k)_{\min}, \qquad v_k = (x'_k)_{\max}. \qquad (11.63)$$

Thus for the admissible solutions in the orthant specified by **s** we have that

$$x'_k \in [(x'_k)_{\min}, (x'_k)_{\max}] = [u_k, v_k]. \qquad (11.64)$$

But by (11.53) we have that

$$x_k = x'_k \quad \text{if} \quad s_k = +1 \quad \text{and} \quad x_k = -x'_k \quad \text{if} \quad s_k = -1. \qquad (11.65)$$

And in the latter case we clearly have that

$$\begin{aligned} (x_k)_{\min} &= -(x'_k)_{\max} = -v_k, \\ (x_k)_{\max} &= -(x'_k)_{\min} = -u_k. \end{aligned} \qquad (11.66)$$

Hence:

The variation of x_k for admissible solutions in the orthant specified by **s** is restricted by $x_k \in [u_k, v_k]$ if   (11.67)
$s_k = +1$, and by $x_k \in [-v_k, -u_k]$ if $s_k = -1$.

Now by way of notation let us call the appropriate interval in (11.67) the *interval of uncertainty in x_k in the orthant specified by* **s** (or briefly, in the **s**-orthant).

Clearly, then, the investigation of the solution set in the **s**-orthant requires obtaining the intervals of uncertainty in this orthant for

x_k, $k = 1, 2, \ldots, n$. And this requires calling the l.p. subroutine $2n$ times.

Summary

On removal of the primes in (11.59) and (11.61), it is clear that the following rules hold for finding the interval of uncertainty in x_k in the s-orthant of an approximate system of equations $\mathbf{Ax} = \mathbf{c}$ with uncertainties $\Delta\mathbf{A} = (\varepsilon_{ij})$, $\Delta\mathbf{c} = (\varepsilon_i)$:

Set up an l.p. problem with the $2n$ constraints

$$\sum_{j=1}^{n} (a_{ij}s_j + \varepsilon_{ij})x_j - x_{i+n} = c_i - \varepsilon_i, \qquad i = 1, 2, \ldots, n,$$

and

$$\sum_{j=1}^{n} (a_{ij}s_j - \varepsilon_{ij})x_j + x_{i+2n} = c_i + \varepsilon_i, \qquad i = 1, 2, \ldots, n,$$

(11.68)

and with nonnegativity conditions

$$x_i \geq 0, \qquad i = 1, 2, \ldots, 3n. \tag{11.69}$$

And with the objective functions specified in (11.62) find the values of x_k, denoting these by u_k and v_k, respectively. Then, the interval of uncertainty in x_k in the s-orthant is given by (11.67), namely,

$$\begin{aligned} x_k &\in [u_k, v_k] && \text{if} \quad s_k = +1, \\ x_k &\in [-v_k, -u_k] && \text{if} \quad s_k = -1. \end{aligned} \tag{11.70}$$

We have thus given the rules for finding the intervals of uncertainty in any required orthant, the l.p. subroutine having to be called $2n$ times. And in the next section we adopt the procedure indicated in this section to determine the intervals of uncertainty in the $\mathbf{x}^0$-orthant of the example in (1.43) and (1.44).

11.7 Example

For our example in (1.43) and (1.44), the vector $\mathbf{x}^0$ is given in (11.32):

$$x_1^0 = 0 \cdot 9730, \qquad x_2^0 = -0 \cdot 9962, \qquad x_3^0 = 2 \cdot 0438. \tag{11.71}$$

Therefore, the orthant in which $\mathbf{x}^0$ lies is specified by $\mathbf{s} = (s_i)$, where

$$s_1 = \text{sign}(x_1^0) = +1, \quad s_2 = \text{sign}(x_2^0) = -1, \quad s_3 = \text{sign}(x_3^0) = +1. \tag{11.72}$$

With these components for **s** the constraints in (11.68) become by (1.43) and (1.44)

$$
\begin{aligned}
0{\cdot}54x_1 - 0{\cdot}85x_2 + 0{\cdot}49x_3 - x_4 && = 0{\cdot}63 \\
0{\cdot}95x_1 + 0{\cdot}48x_2 + 0{\cdot}86x_3 \quad - x_5 && = 3{\cdot}11 \\
0{\cdot}88x_1 - 0{\cdot}54x_2 + 0{\cdot}27x_3 \quad - x_6 && = 0{\cdot}82 \\
0{\cdot}52x_1 - 0{\cdot}87x_2 + 0{\cdot}47x_3 \quad + x_7 && = 0{\cdot}65 \\
0{\cdot}93x_1 + 0{\cdot}46x_2 + 0{\cdot}84x_3 \quad + x_8 && = 3{\cdot}13 \\
0{\cdot}86x_1 - 0{\cdot}56x_2 + 0{\cdot}25x_3 \quad + x_9 && = 0{\cdot}84
\end{aligned}
\tag{11.73}
$$

(compare with (11.38)). And the nonnegativity conditions in (11.69) are

$$
x_i \geq 0, \qquad i = 1, 2, \ldots, 9
\tag{11.74}
$$

(compare with (11.40)).

Then, with the objective functions in (11.41) we obtain

$$
\begin{aligned}
u_1 &= 0{\cdot}8325, & v_1 &= 1{\cdot}1178, \\
u_2 &= 0{\cdot}9276, & v_2 &= 1{\cdot}0672, \\
u_3 &= 1{\cdot}8272, & v_3 &= 2{\cdot}2660.
\end{aligned}
\tag{11.75}
$$

It now follows by (11.70) and (11.72) that

$$
[0{\cdot}8325,\ 1{\cdot}1178],\ [-1{\cdot}0672,\ -0{\cdot}9276],\ \text{and}\ [1{\cdot}8272,\ 2{\cdot}2660]
\tag{11.76}
$$

are the intervals of uncertainty in the three unknowns in the $\mathbf{x}^0$-orthant, the endpoints being the bounds of the closed convex solution subset in the orthant concerned.

And we now note that by (10.131) the intervals in (11.76) are in fact the true intervals of uncertainty of the approximate system of equations in (1.43) and (1.44).

For the orthant clearly contains an insular subset of the solution set, because all the endpoints in (11.76) are finite and none are zero. And, further, the condition $\varepsilon_i \neq 0$, $i = 1, 2, \ldots, n$ in (10.131) holds by (1.44).

Hence it follows by (11.76) that

$$
\begin{aligned}
U_1 &= [0{\cdot}8325,\ 1{\cdot}1178] \\
U_2 &= [-1{\cdot}0672,\ -0{\cdot}9276] \\
U_3 &= [1{\cdot}8272,\ 2{\cdot}2660]
\end{aligned}
\tag{11.77}
$$

(compare with (11.44)).

These, then, are the intervals of uncertainty of the approximate system of equations (1.43) and (1.44) by linear programming (Method IX). And they are the true intervals of uncertainty because they satisfy the conditions for admissible solutions in (10.20) and (10.21).

Next, we deal with the case where the solution set lies in more than one orthant.

11.8 Solution Set Lying In More Than One Orthant

Suppose the solution set does not lie wholly in the $\mathbf{x}^0$-orthant. Then, one could in principle find the intervals of uncertainty in all 2^n orthants, the solution set being the union of the solution subsets in the various orthants.

But this procedure would be very demanding of computation time and we consider the possibility of relaxing the nonnegativity conditions of l.p.

Relaxing Nonnegativity Conditions

When the solution set extends over a number of orthants, the question may be asked whether it is possible to avoid dealing with the orthants separately.

Now, in linear programming it is easy enough to relax the nonnegativity conditions in (11.3) by introducing further variables.

But this approach is not satisfactory in approximate equation theory, the reason being illustrated below by way of the example in (10.23) for $\varepsilon = 0 \cdot 7$ in (10.46) (see Fig. 10.3).

If we relax the nonnegativity conditions, then l.p. will lead us to the region for admissible solutions $ABCD$ in Fig. 10.3. But this is not the entire solution set represented by $GEBCDF$. And the reason for the failure to lead to the entire solution set is that the constraints in the second quadrant are different to those in the first. This is due to the nonlinear nature of the conditions in (10.20) and (10.21) caused by the presence of the terms involving $|x_j|$ in these conditions.

Perhaps another way of explaining why relaxing the nonnegativity conditions cannot lead to the solution set in the example is as follows. It can be proved that l.p. leads to a convex polygon for admissible solutions when the set is bounded. But in our example the solution set represented by $GEBCDF$ is nonconvex (see Fig. 10.3 and (10.57)). Therefore, it cannot be the solution of a single l.p. problem.

Thus, to obtain the solution subset in a particular orthant we have to use the linear constraints appropriate to that orthant. And in general the set of linear constraints obtainable from (10.20) and (10.21) is different for each of the 2^n orthants of solution space.

For it clearly follows by (11.68) that:

The $2n$ constraints corresponding to the conditions for
admissible solutions are functions of the components
of the orthant specifier $\mathbf{s} = (s_i)$ and hence the $2n$ con-   (11.78)
straints are in general different for each of the 2^n
orthants of solution space.

Summarizing, then:

In view of (11.78) relaxing the nonnegativity conditions of linear programming does not in general lead   (11.79)
to the true solution set.

Therefore, to obtain the true bounds on the solution set by linear
programming we have to investigate the solution set in all orthants
containing admissible solutions, avoiding wasteful computation if possible in orthants containing no admissible solutions.

Admissible Orthants

Let us call an orthant containing at least one admissible solution of
solution space an *admissible orthant*. And let us call an orthant containing
no admissible solution an *empty orthant*.

Then, the solution set X is the union of the solution subsets in the
admissible orthants.

Hence, to find the lower endpoint of the interval of uncertainty in
x_k, say, for the approximate system of equations we find the minimum
of the lower endpoints of the orthant intervals of uncertainty in x_k,
there being no need to consider empty orthants. And, similarly, to find
the upper endpoint of the interval of uncertainty in x_k we find the
maximum of the upper endpoints of the orthant intervals of uncertainty
in x_k over the admissible orthants.

And perhaps let us note at this stage that it is possible for identifica-
tion purposes to associate each of the 2^n orthants of solution space with
a unique binary number (or its decimal equivalent). For example, the
orthant specified by $\mathbf{s} = (1, -1, -1, 1)$ can be associated with the
binary number 1001, the binary digit 1 corresponding to a component 1
and the binary digit 0 to a component -1. (And in decimal notation
the orthant in question would be number 9.)

Now, we have already noted that in principle one could find the
intervals of uncertainty in all 2^n orthants and this would lead us to
the intervals of uncertainty of the approximate system of equations.

But it should be clear that all computation should be avoided in
empty orthants. Not only is computation time saved, but the saving

may be such as to make the computation time required feasible which it might otherwise not be.

We therefore consider in the next section the general strategy to be adopted for the solution of the approximate equation problem by linear programming techniques.

11.9 Method IX: Intervals of Uncertainty by Linear Programming

In this section we describe the strategy to be adopted in solving the approximate equation problem by linear programming which will ensure that computation is not carried out with respect to empty orthants. This constitutes our Method IX.

First, let us note that it is not necessary to determine whether the system is critically ill-conditioned or not before applying the procedure. It will be seen that the procedure will either lead to the true intervals of uncertainty or show that the system is critically ill-conditioned.

In particular it will be seen that:

> Method IX will at some stage lead to the discovery that the solution set is unbounded in at least one direction if the system is critically ill-conditioned (see (10.117)). $\qquad$ (11.80)

And in Section 11.10 we discuss the case when information regarding the solution set is nevertheless required although the system is critically ill-conditioned.

But:

> If the solution set is found to be bounded at all stages of the procedure in Method IX then we obtain the true intervals of uncertainty (see (10.115) and (10.104)). $\qquad$ (11.81)

The volume of computation may, however, be very large.

Now, there are two assumptions that we make in Method IX.

First, we assume that

$$\varepsilon_i \neq 0, \qquad i = 1, 2, \ldots, n, \qquad (11.82)$$

in view of the appearance of this condition in (10.117), (10.131), and (10.132). But let us note by (10.129) and (10.134) that the condition is only a sufficient one so that the procedure described may be valid even though the condition in (11.82) does not hold. This is discussed more fully in Section 11.11.

And, second, we assume that the solution set has at least one admissible solution, namely x^0 (see (10.133)).

Now the strategy of Method IX is based essentially on (10.117), (10.104), and (10.115).

It is clear by the above results that if we commence at x^0 then it is possible either to find a continuous path which is unbounded in at least one direction of solution space or otherwise to find a single continuous bounded region containing x^0 and representing the solution set.

Our procedure then is first to determine x^0.

Now, x^0 may have m zero components in which case x^0 lies in 2^m orthants. But for simplicity let us assume that x^0 has no zero components in which case it lies in one orthant only. And the procedure to be adopted when x^0 in fact lies in 2^m orthants will become obvious.

Having found x^0 our second step then is to find the bounds for the solution set in the x^0-orthant by linear programming techniques (see (11.68) to (11.70)).

If any of the endpoints of the orthant intervals of uncertainty so obtained are not bounded then the system is critically ill-conditioned by (10.115) (see also (10.117)). Let us then suppose that all the endpoints are finite.

Now, further, if none of the endpoints are zero then we have an insular subset in the x^0-orthant. We have already dealt with this case in Section 11.7 by way of an example.

Let us therefore suppose that some of the endpoints of the intervals of uncertainty in the x^0-orthant are zero and for simplicity let us first suppose that only one endpoint is zero.

In particular, let us suppose that the lower endpoint of the interval of uncertainty in x_k is zero, i.e.,

$$u_k = 0 \quad \text{if} \quad s_k = +1, \quad v_k = 0 \quad \text{if} \quad s_k = -1 \quad (11.83)$$

(see (11.70)).

Now, since continuity of the solution set is expected both by (10.117) and (10.104), we have a clear indication above that the solution set extends into the orthant with the same specifier s as the x^0-orthant except that it differs in the sign of the kth component. Let us now say that orthants whose specifiers differ only in the sign of one component are *adjacent orthants*.

Then, clearly, we must investigate the solution set in the adjacent orthant indicated by (11.83), i.e., the orthant whose specifier differs in its kth component from that of the x^0-orthant.

Suppose, now, that the solution subset in the adjacent orthant is again bounded.

Then, first, we must note that the upper endpoint now of the interval of uncertainty in x_k in this adjacent orthant is zero. For, by (10.104) and (10.117) we have continuity in passing from the $\mathbf{x}^0$-orthant to the adjacent orthant.

Now, suppose that this is the only zero endpoint in the adjacent orthant.

Then, in fact, the entire solution set lies in the two orthants only and the intervals of uncertainty of the approximate system of equations are the intervals of minimum width containing the orthant intervals of uncertainty. (We may note that the only interval which would contain zero in the above case would be the kth interval of uncertainty.)

Let us now suppose that in investigating the solution set in the $\mathbf{x}^0$-orthant we find that b of the orthant intervals of uncertainty have zero endpoints. Then the solution set lies in at least 2^b orthants because there are 2^b ways of choosing a sequence of b components each of which may be either $+1$ or -1.

Having then investigated the solution set in the $2^b - 1$ orthants other than the $\mathbf{x}^0$-orthant, we may have indications as to which further orthants should be investigated. Further orthants have to be investigated if endpoints are found to be zero in any of the orthant intervals of uncertainty for components of $\mathbf{x}$ other than the b components mentioned earlier.

And if, in fact, $\mathbf{x}^0$ has m zero components instead of none, then the solution set must in the first place be investigated in the 2^m orthants in which $\mathbf{x}^0$ lies and the investigation is then continued from there.

Suppose, now, that a stage is reached where no further orthants are indicated for investigation.

And suppose that:

> All orthant intervals of uncertainty are found to be bounded in the above investigation. (11.84)

It then follows that:

> The approximate system of equations is not critically ill-conditioned by (10.117) and (11.84). (11.85)

For the investigation then clearly shows that no continuous path exists commencing at $\mathbf{x}^0$ which is unbounded in solution space in at least one direction.

But by (10.104) it is clear that:

> The procedure outlined above will ensure that no admissible orthant is excluded from investigation for a non-critically ill-conditioned system of equations. (11.86)

Hence, we have by (10.104) that:

> Subject to the supposition in (11.84) the intervals of
> uncertainty of the approximate system of equations
> are the intervals of minimum width containing the   (11.87)
> orthant intervals of uncertainty over the orthants indi-
> cated for investigation.

Thus, Method IX leads to the true intervals of uncertainty for a
non-critically ill-conditioned system of equations, the assumptions
(11.82) and (10.133) being made.

But it is clear that the volume of computation can be large.

Let us now note that:

> If in the above procedure (Method IX) we find that any
> orthant interval of uncertainty is unbounded then the
> system is critically ill-conditioned by (10.116), this   (11.88)
> result not being subject to the assumption $\varepsilon_i \neq 0$,
> $i = 1, 2, \ldots, n$, in (11.82).

And in the next section we consider the case where the system is
found to be critically ill-conditioned but where information is nevertheless required regarding the solution set.

11.10   Critically Ill-conditioned Systems

Commencing with the $\mathbf{x}^0$-orthant and proceeding to find the orthant
intervals of uncertainty from one admissible orthant to another, as in
Section 11.9, it is clear by (10.116) that the system is critically ill-
conditioned as soon as we find that one component of the solution
vector is unbounded. And this is true whether (11.82) holds or not.

Now, if we are nevertheless interested in the solution set of the
critically ill-conditioned system, then in the first place we proceed as
outlined in Section 11.9 until all indicated orthants are investigated.

But the solution set of a critically ill-conditioned system can be disjoint (see (10.113)). Hence, subsets of the solution set may lie in orthants
other than those investigated.

The question arises: What further orthants are to be investigated?

Of course, one may investigate all the remaining orthants but the
volume of computation will in general become prohibitive.

Equation (10.107), however, gives us a hint as to where we may expect
to find further admissible solutions. As det($\mathbf{A}$) passes through zero we
may well expect x_k in (10.107) to change sign. For example, if for
$n = 3$ the solution subset in the orthant specified by $(1, -1, 1)$, say,

becomes unbounded in the x_3-direction, then one may well expect to find admissible solutions in the orthant specified by $(1, -1, -1)$. And the solution set may be expected to be unbounded in the x_3-direction in this orthant, too. The new orthant could then be the basis for further investigation of the solution set. And let us note that if the solution set becomes unbounded in m directions then this would indicate a further $2^m - 1$ orthants to be investigated.

We have, however, not shown that this procedure will exhaust all admissible orthants. But our interest in this book lies basically in non-critically ill-conditioned systems. So we do not consider critically ill-conditioned systems further. Next we deal with the relaxation of the condition $\varepsilon_i \neq 0$, $i = 1, 2, \ldots, n$, in (11.82).

11.11   The Condition $\varepsilon_i \neq 0$

The condition

$$\varepsilon_i \neq 0, \qquad i = 1, 2, \ldots, n, \tag{11.89}$$

in (11.82) has chiefly a nuisance value (see (10.130), (10.131), (10.132), and (10.117)). First, the condition is generally satisfied. Second, by (10.134) it is a sufficient condition and when it is not satisfied one has the feeling that it is not necessary anyway. But we have shown by way of an example at the end of Section 10.9 that it is not a redundant condition.

Suppose, then, that the condition in (11.82) is not satisfied, and suppose, further, that none of the sufficient conditions for no critical ill-conditioning in (10.135) holds or can be easily applied. Then, if an insular subset X_1 of the solution set exists it may nevertheless be possible to show as follows that the system is not critically ill-conditioned, in which case X_1 is the entire solution set by (10.104).

We introduce uncertainties, preferably small, in those right-hand constants in which the given uncertainties are zero. And we again investigate the solution set. Then, if we find an insular subset X_2 of the new solution set, it follows by (10.130) that the system is not critically ill-conditioned. But the critically ill-conditioned nature of an approximate system of equations depends only on the coefficient matrix $\mathbf{A}$ and the coefficient uncertainty matrix $\Delta\mathbf{A}$ which we have not altered. Hence, the given system of equations is not critically ill-conditioned.

It follows by (10.104) that the insular subset X_1 mentioned earlier represents its entire solution set so that the intervals of uncertainty can be obtained.

But if we find in our investigation of the new solution set when all the $\varepsilon_i \neq 0$ that the solution set is unbounded in any one direction then the system is critically ill-conditioned by (10.115). In this case, then, we cannot say that by (10.104) X_1 represents the entire solution set.

Summarizing, then, it is clear that in particular cases we may be able to prove that the condition in (11.82) is not necessary, which is what we mean by saying that in certain cases the condition in (11.82) can be relaxed.

Investigating the Solution Set Once Only

We now show that in relaxing the condition in (11.82) it may be possible and satisfactory to investigate one solution set only instead of two as previously indicated, namely, using our earlier notation to find X_2 only instead of both X_1 and X_2.

First, let us note that:

> Altering the uncertainties ε_i does not alter the critically ill-conditioned nature of a system of equations; this depends only on $\mathbf{A}$ and $\Delta\mathbf{A}$. $\hspace{2em}$ (11.90)

We have discussed this point earlier in this section.

Second, let us note that:

> The effect of increasing some or all of the ε_i can only be to obtain a solution set containing the original solution set. $\hspace{2em}$ (11.91)

For increasing the ε_i has the effect of decreasing the right-hand sides in (10.20) and increasing the right-hand sides in (10.21).

Now, by (11.90) and (10.130) the system is not critically ill-conditioned if with our previous notation we find an insular subset X_2 of the new solution set. Then, by (10.131) X_2 represents the entire new solution set.

Hence, with X_1 representing the entire original solution set it follows by (11.91) that:

> If an insular subset X_2 is found, then $X_2 \supseteq X_1$. $\hspace{2em}$ (11.92)

Thus, suppose some of the given ε_i are zero. Then, without first investigating the solution set, i.e., without attempting to find X_1, suppose we change the zero uncertainties in the right-hand constants to values just above the noise level of the computation. And suppose we now investigate the solution set.

Then, if we find an insular subset X_2, it follows by (10.130) and (11.90) that the original system is not critically ill-conditioned.

Hence, if we accept the intervals of uncertainty obtained from X_2 as of satisfactory width, then we need only investigate one solution set, instead of two as previously indicated when the condition in (11.82) does not hold.

11.12 Conclusion

We have shown that the conditions for admissible solutions in (10.20) and (10.21) can be transformed to standard linear programming form to obtain the intervals of uncertainty. And we emphasize that the intervals of uncertainty found by the linear programming approach are always the true intervals of uncertainty.

We have seen that in some cases only the linear programming approach, i.e., only using the conditions for admissible solutions in (10.20) and (10.21), will lead to the true intervals of uncertainty (see Sections 10.5 and 10.6, and Table 10.2). And, indeed, in some situations the l.p. approach is applicable when all the other methods fail to give any estimates of the intervals of uncertainty (see Section 10.4).

Clearly, the l.p. approach can be very useful in certain situations in approximate equation theory.

Now, when the solution set lies entirely in the $\mathbf{x}^0$-orthant, then the l.p. approach involves calling the l.p. subroutine $2n$ times. The volume of computation involved is of the same order as for Method III and perhaps cannot be regarded as excessive.

But when the solution set is spread over many orthants, the volume of computation can become very large, and indeed prohibitive, the total number of orthants in an n-dimensional problem being 2^n. Under these circumstances, we must be satisfied with the results of methods leading to intervals containing the true intervals of uncertainty. And Method VI or VIII gives the intervals of smallest width containing the true intervals of uncertainty.

In conclusion, then, let us say that the intervals of uncertainty obtained by the l.p. approach are always the true intervals. Further, in certain cases the only possible method of obtaining the true intervals of uncertainty is by the l.p. approach, and this may even be the case when all other methods fail to give any estimates of the uncertainties.

CHAPTER 12

A Statistical Approach

12.1 Introduction

We have not so far considered the criteria for specifying the uncertainties ε_{ij} and ε_i in the coefficients and right-hand constants, the true system of equations being denoted by $\mathbf{A^*x^*} = \mathbf{c^*}$ and the approximate system of equations by $\mathbf{Ax} = \mathbf{c}$.

If the values of the coefficients and right-hand constants are obtained by measurement, let us suppose that a series of values are found for each coefficient and constant. And suppose we now assume that the coefficient a_{ij}^*, say, lies somewhere in the interval defined by the smallest and largest measurements for it. Then we could take half the length of the interval containing the measurements as the uncertainty ε_{ij} and the midpoint of the interval as the approximate value a_{ij}.

But the above approach is primitive. For, clearly, the uncertainties ε_{ij} and ε_i specified in this way might become larger if we were to take more measurements.

Let us rather assume that the measured values of the a_{ij}^* and c_i^* obey some distribution curve and let us adopt a statistical approach.

A distribution curve that frequently occurs in practice is the *normal distribution curve*, and it is often assumed that the measured values for a quantity obey this curve. (We may point out that it is possible to test whether measurements in fact conform to the normal distribution curve but we do not deal with this topic here.) And, further, we require the tacit assumption that we have made throughout this book that the measured values of any of the a_{ij}^* or c_i^* are independent of the measurements for all the other a_{ij}^* and c_i^*.

Then, we can find statistical estimates of the ill-conditioning factor and of the uncertainties in the unknowns for given confidence limits. And it will be seen that the results largely hold when the measured values of the a_{ij}^* and c_i^* are not normally distributed.

In adopting the statistical approach we have to decide on acceptable confidence limits, i.e., with what probability we are prepared to accept that our estimate of the ill-conditioning factor and our estimates of the uncertainties in the unknowns are not too small.

Then, in practice, adopting the statistical approach leads to smaller values of the ill-conditioning factor and of the uncertainties in the unknowns than we would otherwise obtain. This, then, is the importance of our present approach.

12.2   The Normal Distribution Curve

Let us denote the measured values of one of the coefficients a_{ij}^* or one of the right-hand constants c_i^* by

$$m_1, m_2, \ldots, m_h, \tag{12.1}$$

it being supposed that h measurements are taken. And let us denote the mean of these h measurements by

$$m = (m_1 + m_2 + \ldots + m_h)/h. \tag{12.2}$$

Then, m is called the *sample mean*. For in statistics the term *population* refers to the totality of measurements for a quantity, the measurements actually taken being regarded as a *sample* chosen from the population.

Let us now suppose that to obtain the population mean, denoted by μ, we let h approach infinity in (12.2). Thus,

$$m \longrightarrow \mu \quad \text{as} \quad h \longrightarrow \infty. \tag{12.3}$$

While we develop our theory in terms of the population measures, we approximate the population measures by the corresponding sample measures when this is convenient and appropriate. And, of course, the larger the number of measurements made for a quantity, i.e., the larger h, the better the approximation.

Now, a measure of the spread of the h measurements is given by s defined by

$$s^2 = \frac{(m - m_1)^2 + (m - m_2)^2 + \ldots + (m - m_h)^2}{h - 1}, \tag{12.4}$$

and let us not concern ourselves here with the reason for having $h - 1$ in the denominator instead of the h one might expect.

Then, s is called the *sample standard deviation* and s^2 is called the *sample variance*. As h approaches infinity our sample becomes the population, the *population standard deviation* and the *population variance* being denoted by σ and σ^2, respectively.

We are now in a position to give the equation of the *normal distribution curve* sketched in Fig. 12.1.

But first let us note that a function $y = \phi(x)$ is said to represent a distribution curve if the probability that a measurement lies between x and $x + \delta x$ is $\phi(x)\,\delta x$, i.e., the area under the curve between x and $x + \delta x$. And the total area under the curve $y = \phi(x)$ must clearly be

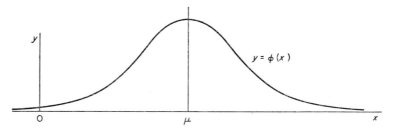

FIG. 12.1　The normal distribution curve

unity because the probability that a measurement has some value is unity. Thus,

$$\int_{-\infty}^{\infty} \phi(x)\,dx = 1. \tag{12.5}$$

The equation of the so-called normal distribution curve for a population with mean μ and standard deviation σ is

$$y = \frac{1}{\sigma(2\pi)^{\frac{1}{2}}} e^{-\frac{1}{2}(x-u)^2/\sigma^2} = \phi(x) \tag{12.6}$$

where $\pi = 3 \cdot 14159 \ldots$ and $e = 2 \cdot 71828 \ldots$. And it can be shown that (12.5) is satisfied for the normal distribution curve in (12.6).

Now, we do not generally know the population measures μ and σ but only the sample measures m and s. And, while in some situations it is sufficiently good in practice to approximate the population measures by the sample measures, this aspect may require careful consideration. And this is especially the case when the number of elements in the sample is small, i.e., when h in (12.1) is small.

Our development of the theory nevertheless proceeds in terms of population measures.

Now, the importance of the value of the standard deviation can be seen from the following properties. About $68 \cdot 27$ per cent of the area under the normal distribution curve lies within a distance σ along the x-axis from the mean position μ, $95 \cdot 45$ per cent of the area within a distance 2σ, and 99·73 per cent within a distance 3σ (see Fig. 12.2). Thus the probability that a measured value lies outside the interval

$$[\mu - 3\sigma, \mu + 3\sigma] \tag{12.7}$$

is only $(100 - 99\cdot73)$ per cent $= 0\cdot27$ per cent.

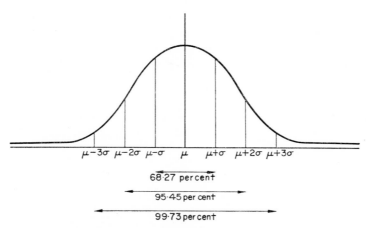

FIG. 12.2   Confidence limits corresponding to σ, 2σ, and 3σ

In general, the probability of a measured value lying in a given interval can be determined if the distribution curve is given. But in the statistical approach to approximate equation analysis, we are interested in the probability of the 'true value' lying in the interval of uncertainty. And in setting up the approximate equation problem we wish to determine intervals of uncertainty in the coefficients and right-hand constants such that these will contain their respective true values for given probabilities.

Let us, then, consider the term 'true value' more closely for one of the coefficients or right-hand constants.

Where a number of measurements are made for a quantity, the true value is taken as the population mean, the measurements being frequently normally distributed about the population mean. In Section 12.10 we deal with the case where the true value of a coefficient or right-hand constant is to be interpreted in this way.

But for the moment let us consider a different practical situation.

Let us suppose that the coefficient a_{11}, to be specific, represents the stiffness of a spring used in a certain situation, the spring being chosen at random from a large number of almost identical springs. This could be the situation in some manufacturing process. Suppose also that it is not convenient to measure the stiffness at the time the spring is chosen. And suppose that the distribution curve for the measurements of the stiffnesses of all the springs is known, the errors of measurement being able to be ignored, and one measurement being taken for each spring. (In practice the whole population would not be observed but a random sample would be taken.)

Then, it is easy to determine the probability that the true value of a spring chosen at random lies in a particular interval. In fact, the probability of the true value lying in a given interval is the same as that of finding measured values in the interval.

Thus:

> The probability that the true value of a quantity in the above situation lies in an interval between x and $x + \delta x$ is $\phi(x)\, \delta x$ where $y = \phi(x)$ is the distribution curve for the measurements. (12.8)

And in brief we may say that:

> The distribution curve of the true value of a quantity in the above situation is that of the measurements themselves. (12.9)

We may emphasize that (12.8) and (12.9) hold for any distribution curve. And in particular, (12.8) and (12.9) hold for the normal distribution curve.

We can therefore say, for example, that the probability that the true value lies outside the interval in (12.7) is only $(100 - 99 \cdot 73)$ per cent $= 0 \cdot 27$ per cent. Or, put in another way, we may say that the true value lies in the interval in (12.7) with a *confidence limit* of $99 \cdot 73$ per cent.

Clearly, then, if the standard deviation is small the length of the interval of uncertainty for a given confidence limit is small.

In developing the statistical approach, we shall keep the practical situation outlined in this section in mind until the end of Section 12.8.

12.3   The Uncertainties and the Standard Deviations

Let us take the uncertainty ε of a quantity whose true value is t, say, as some constant p times the population standard deviation of the measurements, i.e., we take

$$\varepsilon = p\sigma. \qquad (12.10)$$

The choice of p is left to us, and on it depends the confidence limit.

Thus, if μ is the population mean for the measurements then clearly the probability that the true value t lies in the interval

$$[\mu - \varepsilon, \mu + \varepsilon] \qquad (12.11)$$

depends on p in (12.10).

If the measurements are normally distributed then a choice of $p = 3$ implies a confidence limit of 99·73 per cent that the true value lies in the interval in (12.11). In other words, a choice of $p = 3$ implies a confidence limit of 99·73 per cent that the uncertainty $\varepsilon = p\sigma$ is not too small for the interval in (12.11) to contain the true value. And for $p = 4$ the confidence limit is very nearly 100·00 per cent.

Now, suppose we take

$$\varepsilon_{ij} = p\sigma_{ij}, \qquad \varepsilon_i = p\sigma_i, \qquad i, j = 1, 2, \ldots, n \qquad (12.12)$$

where the σ_{ij} and the σ_i are the population standard deviations of the a_{ij}^* and c_i^*. (By (12.9) we may regard statistical measures as referring to the distribution curve either of the measurements or of the true value.) And let us suppose that the population means of the measurements give our so-called approximate values a_{ij} and c_i, although in practice these values are taken to be the sample means.

Then the confidence limits for the following statements to be true

$$a_{ij}^* \in [a_{ij} - \varepsilon_{ij}, a_{ij} + \varepsilon_{ij}]$$
$$c_i^* \in [c_i - \varepsilon_i, c_i + \varepsilon_i], \qquad i, j = 1, 2, \ldots, n \qquad (12.13)$$

clearly depend on our choice of p in (12.12) and, of course, on the distribution curves of the a_{ij}^* and c_i^*. And when the distribution curves are normal, a choice of $p = 3$ or $p = 4$ would appear appropriate in many circumstances. Indeed, this choice of p will later be seen to appear appropriate from the role p plays in (12.34) to (12.36) and in (12.51).

In Section 12.4 we first give some results from statistics before dealing with the statistical ill-conditioning factor in Section 12.5.

12.4 Some Results from Statistics

Change of Variable

We now consider the effect of introducing a change of variable

$$z_i = m_i - a, \qquad i = 1, 2, \ldots, h, \qquad (12.14)$$

in (12.2) and (12.4) where a is a constant.

It will be seen that the mean is reduced by a but that the standard deviation is unaltered.

For, by (12.2) the mean of the z_i is

$$\frac{(m_1 - a) + (m_2 - a) + \ldots + (m_h - a)}{h}$$

$$= \frac{m_1 + m_2 + \ldots + m_h}{h} - \frac{ha}{h} = m - a \tag{12.15}$$

(see (12.2)).

Thus, the sample mean of the distribution is reduced by a.

And, since the result in (12.15) holds for all values of h, it follows that the population mean, too, is reduced by a.

Thus:

> The change of variable in (12.14) reduces the population mean and all sample means by a.   (12.16)

The above result is indeed expected. For the change of variable in (12.14) merely displaces the distribution curve to the left through a distance a. This displacement clearly results in the mean being reduced by a, as we have shown.

But, clearly, all measures and properties associated with the shape of a distribution curve are unaltered by a displacement of the curve.

Thus:

> The change of variable in (12.14) does not change the shape of the distribution curve and hence leaves all measures and properties associated with the shape of the distribution curve unaltered. In particular, there is no change in the standard deviation, and, further, if the m_i are normally distributed then the z_i must also be, and vice versa.   (12.17)

We may note that the result that the standard deviation is unaltered can be easily checked by (12.4), (12.14), and (12.15). For, clearly, sample standard deviation of the z_i

$$= \left(\frac{(m - a - z_1)^2 + (m - a - z_2)^2 + \ldots + (m - a - z_h)^2}{h - 1}\right)^{\frac{1}{2}}$$

$$= \left(\frac{(m - m_1)^2 + (m - m_2)^2 + \ldots + (m - m_h)^2}{h - 1}\right)^{\frac{1}{2}}$$

$=$ sample standard deviation of the m_i.

And this result holds, as $h \to \infty$, so that it holds for the population standard deviation.

Let us now use the above results in our approximate equation theory.

First, let us note that the mean of the distribution curve of a_{ij}^* $(i, j$ fixed) is a_{ij}, the population mean of the measurements for a_{ij}^*, because by (12.9) the two distributions must have identical means, the practical situation being considered being the one outlined towards the end of Section 12.2.

Then, let us consider the new variable δa_{ij}, defined by

$$\delta a_{ij} = a_{ij}^* - a_{ij}, \qquad (12.18)$$

a_{ij} clearly being a constant.

It then follows by (12.9), (12.16), and (12.17) that:

> The variable $\delta a_{ij} = a_{ij}^* - a_{ij}$ is distributed with zero mean and standard deviation σ_{ij} (see (12.12)), the shape of the distribution curve of δa_{ij} being the same as of a_{ij}^* or of the measurements for a_{ij}^*. And in particular, δa_{ij} is normally distributed if a_{ij}^* is normally distributed, and vice versa. (12.19)

And similarly for the variable

$$\delta c_i = c_i^* - c_i \qquad (12.20)$$

(i fixed) we have by (12.9), (12.16), and (12.17) that:

> The variable $\delta c_i = c_i^* - c_i$ is distributed with zero mean and with standard deviation σ_i (see (12.12)). And, in particular, δc_i is normally distributed if c_i^* is normally distributed, and vice versa. (12.21)

The result in (12.19), of course, holds for each set of values of i, j $(i, j = 1, 2, \ldots, n)$ and the result in (12.21) for $i = 1, 2, \ldots, n$.

The Sum of Independent Random Variables

First, we shall indicate what we understand by the term 'independent random variables' in our context, and then we shall state the results giving their sum.

We have assumed tacitly throughout that any measurement for one of the coefficients or right-hand constants is in no way affected by any of the previous measurements for this quantity, nor in any way affects any subsequent measurements for this quantity (although the distribution curve for the measurements may be known). If this assumption is valid then the measurements for the quantity are elements of a *random variable*.

If the measurements are for a_{ij}^*, say, for some fixed i and j, then it follows by (12.9) that a_{ij}^* itself can be regarded as a random variable. And it then follows that $\delta a_{ij} = a_{ij}^* - a_{ij}$ is a random variable (see (12.19)). Similarly, if the measurements are for c_i^* then the variable δc_i in (12.20) is a random variable.

And, if we assume that the measurements for any one of the coefficients or right-hand constants in no way affects the measurements for any other of the coefficients or right-hand constants then the random variables in question are *independent*.

Hence:

> The δa_{ij} $(i, j = 1, 2, \ldots, n)$ and the δc_i $(i = 1, 2, \ldots, n)$ form a set of $n^2 + n$ independent random variables. $\qquad$ (12.22)

Let us now consider the effect of multiplying a random variable r by a scalar a, i.e., multiplying each value of r by a, and let us denote the new random variable so obtained by y. The following result then clearly follows by (12.2), (12.4), and (12.6):

> If r is a random variable with mean μ and standard deviation σ and if a is a scalar then $y = ar$ is a random variable with mean $a\mu$ and standard deviation $|a|\sigma$, and y is normally distributed if r is normally distributed. $\qquad$ (12.23)

This result is in fact a particular case of the following more general result (see reference 1, pp. 48 and 49):

If

(a) $r_1, r_2, \ldots, r_n$ are independent random variables having (population) means $\mu_1, \mu_2, \ldots, \mu_n$ and (population) variances $\sigma_1^2, \sigma_2^2, \ldots, \sigma_n^2$, respectively,

(b) $a_1, a_2, \ldots, a_n$ are constants, and

(c) $y = a_1r_1 + a_2r_2 + \ldots + a_nr_n$, $\qquad$ (12.24)

then it is well known from probability theory that y is a random variable with the following three properties:

(a) mean of y, $\mu_y = a_1\mu_1 + a_2\mu_2 + \ldots + a_n\mu_n$;

(b) variance of y, $\sigma_y^2 = a_1^2\sigma_1^2 + a_2^2\sigma_2^2 + \ldots + a_n^2\sigma_n^2$; $\qquad$ (12.25)

(c) and if $r_1, r_2, \ldots, r_n$ are normally distributed then y is also normally distributed (with mean μ_y given by (a) and variance σ_y^2 given by (b)).

Thus, if $y = a_1r_1 + a_2r_2 + \ldots + a_nr_n$ is evaluated for each set of values obtained for $r_1, r_2, \ldots, r_n$ then y is a random variable with the properties in (12.25). And, in particular, we note that y is normally distributed if the r_i are.

But suppose $r_1, r_2, \ldots, r_n$ are not normally distributed. Then the *central limit theorem* comes to our aid.

For the case when the n variables $r_1, r_2, \ldots, r_n$ are *identically distributed* independent random variables (i.e., the distribution curves have the same form) the theorem states that the distribution of y approaches the normal distribution for n large. And further the theorem states that the sum of a large number of independent random variables will be approximately normally distributed regardless of the distributions of the individual random variables provided the variances are finite.

It is for this reason that the errors of measurement, for example, are frequently assumed to be normally distributed, each error being usually composed of the sum of many small individual components. And it would then follow that the measurements themselves are normally distributed. Similarly if the aim is to produce a number of identical objects the errors during manufacture can be expected to lead to any property associated with these objects to be normally distributed.

We now note that the following result holds by (12.24), (12.25), and the central limit theorem when the random variables all have zero means:

> Given a number of independent random variables, $r_1, r_2, \ldots, r_n$, each with zero mean and finite variance, then $y = a_1 r_1 + a_2 r_2 + \ldots + a_n r_n$ in (c) of (12.24) is a random variable with zero mean and variance given by (b) of (12.25); the random variable y is normally distributed if the given independent variables are normally distributed, but otherwise y can only be said to approach the normal distribution.   (12.26)

12.5   The Statistical Ill-conditioning Factor ϕ_s

We shall define a statistical ill-conditioning factor ϕ_s such that the condition for no critical ill-conditioning

$$\phi_s < 1$$

will be subject to a confidence limit specified by p in (12.12).

Denoting the determinant of the true coefficient matrix $\mathbf{A}^* = (a_{ij}^*)$ by a^* and of the approximate coefficient matrix $\mathbf{A} = (a_{ij})$ by a, i.e., $a^* = \det(\mathbf{A}^*)$ and $a = \det(\mathbf{A})$, the increment $\delta a = a^* - a$ in the value of the determinant due to changes in the coefficients is given by

$$\delta a = \sum_{i=1}^{n} \sum_{j=1}^{n} \frac{\partial a^*}{\partial a_{ij}^*} \delta a_{ij} \qquad (12.27)$$

to the first order of small quantities (see (1.19)).

Now taking the partial derivatives in (12.27) at $\mathbf{A}^* = \mathbf{A}$ we have by (1.34) that

$$\frac{\partial a^*}{\partial a_{ij}^*} = ab_{ji}, \qquad i, j = 1, 2, \ldots, n. \qquad (12.28)$$

Hence by (12.28) and (12.27) we have

$$\delta a = \sum_{i=1}^{n} \sum_{j=1}^{n} ab_{ji} \, \delta a_{ij}. \qquad (12.29)$$

Now since by (12.22) and (12.19) the δa_{ij} are independent random variables with zero means and since the ab_{ji} are constants it follows by (12.24), (12.25), and (12.26) that:

> The increment δa in the determinant is a random variable with zero mean which is normally distributed if the δa_{ij} are each normally distributed, but otherwise can only be said to approach the normal distribution. $\qquad (12.30)$

And using the same type of arguments as leading to (12.19) we now have that:

> The shape of the distribution curve of the true value of the determinant a^* is the same as for δa, and in particular a^* and δa have the same standard deviation σ_a, say. Further, since δa in (12.27) has zero mean by (12.30), it follows that the population mean of the distribution curve for a^* is a, the determinant of the approximate coefficient matrix $\mathbf{A} = (a_{ij})$. $\qquad (12.31)$

Hence, by (12.29), (12.25), and (12.19)

$$\sigma_a^2 = a^2 \sum_{i=1}^{n} \sum_{j=1}^{n} b_{ji}^2 \sigma_{ij}^2. \qquad (12.32)$$

Hence, by (12.12) we have

$$\sigma_a^2 = a^2 \sum_{i=1}^{n} \sum_{j=1}^{n} b_{ji}^2 \varepsilon_{ij}^2 / p^2 \qquad (12.33)$$

so that

$$p\sigma_a = |a| \left(\sum_{i=1}^{n} \sum_{j=1}^{n} b_{ji}^2 \varepsilon_{ij}^2 \right)^{\frac{1}{2}}. \tag{12.34}$$

Now $p\sigma_a$ represents a change in value of the determinant which is not exceeded with a confidence limit specified by p (and of course by the form of the distribution curve, but see (12.30)). Thus the confidence limit for a^* to lie in the interval

$$[a - p\sigma_a, a + p\sigma_a] \tag{12.35}$$

is specified by p, a being the value of the determinant of the approximate coefficient matrix $\mathbf{A} = (a_{ij})$ (see (12.31)). And if a^* is in fact normally distributed then for $p = 3$ the confidence limit that a^* lies in the interval in (12.35) is 99·73 per cent.

Now the condition that the interval in (12.35) does not include zero is clearly

$$p\sigma_a < a \quad \text{if} \quad a > 0 \quad \text{and} \quad p\sigma_a < -a \quad \text{if} \quad a < 0,$$

i.e.,

$$\frac{p\sigma_a}{|a|} < 1. \tag{12.36}$$

This, then, is the condition for no critical ill-conditioning with a confidence limit specified by p, a system of equations being said to be critically ill-conditioned if the determinant of the coefficient matrix can become zero within the limits of the uncertainties in the coefficients.

Hence by (12.34) and (12.36) the condition for no critical ill-conditioning with a confidence limit specified by p is

$$\phi_s < 1 \tag{12.37}$$

where

$$\phi_s = \left(\sum_{i=1}^{n} \sum_{j=1}^{n} b_{ji}^2 \varepsilon_{ij}^2 \right)^{\frac{1}{2}}. \tag{12.38}$$

And we call ϕ_s the *statistical ill-conditioning factor*.

In determining the confidence limit for (12.37) to hold it should be noted that we may safely assume a^* to be normally distributed because the approach to the normal distribution referred to in (12.30) is in general very rapid as the number of independent random variables increases (see also (12.31)).

Next, we note that

$$\phi_s \leq \phi, \tag{12.39}$$

a result which follows directly from the definitions. For by (12.38) and (1.35) we have that

$$\phi_s = \left(\sum_{i=1}^{n} \sum_{j=1}^{n} b_{ji}^2 \varepsilon_{ij}^2 \right)^{\frac{1}{2}} \leq \sum_{i=1}^{n} \sum_{j=1}^{n} |b_{ji}| \, \varepsilon_{ij} = \phi. \tag{12.40}$$

And it is the relation in (12.39) that indicates the importance of the statistical approach to the ill-conditioning factor. For with the $\varepsilon_{ij} = p\sigma_{ij}$ defined as in (12.12) we may find in certain cases in which (1.29) is not satisfied, i.e., when ϕ is equal to or greater than unity, that the condition $\phi_s < 1$ in (12.37) is nevertheless satisfied, it being assumed that the confidence limit specified by p is regarded as adequate. And where necessary we can determine the largest value of p and hence the largest confidence limit for $\phi_s < 1$ in (12.37) still to hold. For writing (12.38) as

$$\phi_s = p \left(\sum_{i=1}^{n} \sum_{j=1}^{n} b_{ji}^2 \sigma_{ij}^2 \right)^{\frac{1}{2}} \tag{12.41}$$

(see (12.12)) it is clear that the value of ϕ_s is directly proportional to the value of p.

Let us now note that for the special case when all the $\varepsilon_{ij} = \varepsilon$ the condition for no critical ill-conditioning in (12.37) and (12.38) becomes

$$\phi_s = \left(\sum_{i=1}^{n} \sum_{j=1}^{n} b_{ij}^2 \right)^{\frac{1}{2}} \varepsilon < 1, \tag{12.42}$$

the confidence limit being specified by the value of p used in (12.12).

Example

For our numerical example in (1.43) and (1.44) we have by (1.46) and (12.42) that

$$\phi_s = 3 \cdot 483 \times 0 \cdot 01 = 0 \cdot 03483. \tag{12.43}$$

For comparison we have by (1.48) that

$$\phi = 0 \cdot 08611 \tag{12.44}$$

so that (12.39) is clearly satisfied.

Suppose that in our numerical example p was taken equal to 3, this leading to $\phi_s = 0 \cdot 03483$ in (12.43). Then clearly the condition $\phi_s < 1$ in (12.37) would be satisfied provided p is chosen such that

$$p < 3/0 \cdot 03483 = 3 \times 28 \cdot 7 = 86 \cdot 1.$$

For by (12.42) the value of ϕ_s is directly proportional to value of ε, and by (12.12) this is directly proportional to the value of p (or see (12.41)).

Clearly then our example is not critically ill-conditioned with a confidence limit specified by a value of p which may be as high as, say, 28 times the value actually used in (12.12).

12.6 The Standard Deviations of the Solution

In this section we derive expressions for the standard deviations

$$\sigma_{x_1}, \sigma_{x_2}, \ldots, \sigma_{x_n} \tag{12.45}$$

in the unknowns $x_1^*, x_2^*, \ldots, x_n^*$.

To do this we use the expressions in (4.13) giving the δx_k approximately as in Method II:

$$\delta x_k = -\sum_{i=1}^{n} \sum_{j=1}^{n} x_j b_{ki} \, \delta a_{ij} + \sum_{i=1}^{n} b_{ki} \, \delta c_i, \qquad k = 1, 2, \ldots, n, \tag{12.46}$$

where

$$\delta x_k = x_k^* - x_k, \qquad k = 1, 2, \ldots, n, \tag{12.47}$$

(see (1.18)).

Then, since the δa_{ij} and δc_i are independent random variables with zero means (see (12.19), (12.21), and (12.22)), it follows by (12.26) and (12.46) that δx_k, for each of $k = 1, 2, \ldots, n$, is a random variable with zero mean and is normally distributed if the δa_{ij} and δc_i are normally distributed. Otherwise each δx_k nevertheless approaches the normal distribution.

And it follows by (12.47) that the distribution curves of the δx_k and the x_k^* are identical in shape, in particular the standard deviations of the δx_k being those of the x_k^*, namely, $\sigma_{x_1}, \sigma_{x_2}, \ldots, \sigma_{x_n}$.

Thus:

> Each δx_k ($k = 1, 2, \ldots, n$) has zero mean and is either normally distributed or approaches the normal distribution, it being assumed in this treatment that it is sufficiently good to regard each δx_k as normally distributed. And the x_k^* have distribution curves of identical shape to those of the δx_k, the standard deviations of the x_k^* and δx_k being denoted by $\sigma_{x_1}, \sigma_{x_2}, \ldots, \sigma_{x_n}$. $\tag{12.48}$

It now follows by (12.46) and (b) of (12.25) that

$$\sigma_{x_k} = \sqrt{\left(\sum_{i=1}^{n} \sum_{j=1}^{n} (x_j b_{ki})^2 \sigma_{ij}^2 + \sum_{i=1}^{n} b_{ki}^2 \sigma_i^2 \right)}, \qquad k = 1, 2, \ldots, n \tag{12.49}$$

(see (12.19) and (12.21)).

The standard deviations of the x_k^* are thus given by (12.49).

In passing let us note that since the δx_k as approximated by (12.46) have zero means, the approximation in (12.46) implies that the x_k $(k = 1, 2, \ldots, n)$ in (12.47) represent the population means of the x_k^*. But we may recall that $x = (x_k)$ was in fact originally taken to be the solution of our approximate system of equations $\mathbf{Ax} = \mathbf{c}$ where $\mathbf{A}$ and $\mathbf{c}$ are given by the population means of the measurements. Hence we may note that:

> The population means of the x_k^* can be approximated by the components of the solution $\mathbf{x} = (x_k)$ of our approximate system of equations $\mathbf{Ax} = \mathbf{c}$ where $\mathbf{A}$ and   (12.50) $\mathbf{c}$ are given by the population means of the measurements.

12.7 The Statistical Uncertainties

We define the *statistical uncertainties* Δx_k^S by

$$\Delta x_k^S = p\sigma_{x_k}, \qquad k = 1, 2, \ldots, n, \qquad (12.51)$$

the superscript S standing for *statistical*.

It thus follows by (12.12) that the Δx_k^S represent the same multiples of the standard deviations in the x_k^* as do the ε_{ij} and ε_i of the standard deviations in the a_{ij}^* (or δa_{ij}) and in the c_i^* (or δc_i).

We now find expressions for the Δx_k^S.

By (12.12) and (12.49) we have that

$$p^2\sigma_{x_k}^2 = \sum_{i=1}^{n}\sum_{j=1}^{n}(x_j b_{ki}\varepsilon_{ij})^2 + \sum_{i=1}^{n}(b_{ki}\varepsilon_i)^2, \qquad k = 1, 2, \ldots, n. \quad (12.52)$$

Hence, by (12.51)

$$\Delta x_k^S = \left(\sum_{i=1}^{n}\sum_{j=1}^{n}(x_j b_{ki}\varepsilon_{ij})^2 + \sum_{i=1}^{n}(b_{ki}\varepsilon_i)^2\right)^{\frac{1}{2}}, \qquad k = 1, 2, \ldots, n. \quad (12.53)$$

With the above expressions for the Δx_k^S, the confidence limit with which the following statement holds

$$x_k^* \in [x_k - \Delta x_k^S, \; x_k + \Delta x_k^S], \qquad k = 1, 2, \ldots, n \qquad (12.54)$$

is determined by the value of p in (12.12), the x_k^* being normally distributed by (12.48). For example, for $p = 3$ in (12.12) we have that (12.54) holds with a confidence limit of 99·73 per cent. And when each of the δa_{ij} and δc_i is itself normally distributed then (12.54) holds with the same confidence limits as do the statements in (12.13).

(We may not in fact know any of the population means. In practice we take the x_k in (12.54) and the x_j in (12.49) and (12.53) as the elements of the solution vector of $\mathbf{Ax} = \mathbf{c}$ where $\mathbf{A}$ and $\mathbf{c}$ may have to be approximated by the sample means (see (12.50)). And the b_{ki} are then the elements of $\mathbf{B} = \mathbf{A}^{-1}$.)

Let us now note than in the particular case when

$$\varepsilon_{ij} = \varepsilon = \varepsilon_i, \qquad i, j = 1, 2, \ldots, n \tag{12.55}$$

the form which Δx_k^{S} takes is given by

$$\Delta x_k^{\mathrm{S}} = \left(\sum_{i=1}^{n} \sum_{j=1}^{n} (x_j b_{ki})^2 + \sum_{i=1}^{n} (b_{ki})^2 \right)^{\frac{1}{2}} \varepsilon, \qquad k = 1, 2, \ldots, n,$$

i.e.,

$$\Delta x_k^{\mathrm{S}} = \left((1 + \sum_{j=1}^{n} x_j^2)(\sum_{i=1}^{n} b_{ki}^2) \right)^{\frac{1}{2}} \varepsilon, \qquad k = 1, 2, \ldots, n. \tag{12.56}$$

Now, in concluding this section we recall that the Δx_k^{II} and the Δx_k^{S} were derived from the same approximation (see (4.13) and (12.46)). And we recall that the Δx_k^{II} are close to the Δx_k, especially in well-conditioned systems of equations, by which we mean systems for which $\|\mathbf{C}\|_{\mathrm{I}} \ll 1$ (consider (1.53) and (7.26), for example). Hence, let us regard the values of the Δx_k^{S} as close enough for the confidence limit specified by p in all but exceptional cases, the exceptional cases being those for which ϕ_s is less than but very close to unity (see (12.37)).

12.8 Example

For our numerical example in (1.43) for an uncertainty $\varepsilon = 0 \cdot 01$ we have by (12.56) that

$$\begin{aligned}
\Delta x_1^{\mathrm{S}} &= \varepsilon \times ((1 + x_1^2 + x_2^2 + x_3^2)(b_{11}^2 + b_{12}^2 + b_{13}^2))^{\frac{1}{2}} \\
&= 0 \cdot 01 \times (7 \cdot 11626 \times 4 \cdot 06559)^{\frac{1}{2}} \\
&= 0 \cdot 01 \times (28 \cdot 9318)^{\frac{1}{2}} = 0 \cdot 01 \times 5 \cdot 379 = 0 \cdot 05379.
\end{aligned}$$

Similarly,

$$\begin{aligned}
\Delta x_2^{\mathrm{S}} &= \varepsilon \times ((1 + x_1^2 + x_2^2 + x_3^2)(b_{21}^2 + b_{22}^2 + b_{23}^2))^{\frac{1}{2}} \\
&= 0 \cdot 01 \times (7 \cdot 11626 \times 1 \cdot 03989)^{\frac{1}{2}} = 0 \cdot 02720
\end{aligned}$$

and

$$\begin{aligned}
\Delta x_3^{\mathrm{S}} &= \varepsilon \times ((1 + x_1^2 + x_2^2 + x_3^2)(b_{31}^2 + b_{32}^2 + b_{33}^2))^{\frac{1}{2}} \\
&= 0 \cdot 01 \times (7 \cdot 11626 \times 7 \cdot 02315)^{\frac{1}{2}} = 0 \cdot 07070.
\end{aligned}$$

Thus,

$$\Delta x_1^S = 0 \cdot 05379, \qquad \Delta x_2^S = 0 \cdot 02720, \qquad \Delta x_3^S = 0 \cdot 07070. \qquad (12.57)$$

Let us now compare these results with the Δx_k^{II}, namely,

$$\Delta x_1^{II} = 0 \cdot 1426, \qquad \Delta x_2^{II} = 0 \cdot 0697, \qquad \Delta x_3^{II} = 0 \cdot 2193$$

(see (4.35)), because the approximation in (12.46) was used in developing both methods. We then have

$$\Delta x_1^S = 0 \cdot 377\, \Delta x_1^{II}, \qquad \Delta x_2^S = 0 \cdot 390\, \Delta x_2^{II}, \qquad \Delta x_3^S = 0 \cdot 322\, \Delta x_3^{II}. \tag{12.58}$$

Clearly, the statistical approach gives smaller and more meaningful values for the uncertainties in the unknowns.

And it is easy enough to determine the uncertainties if a different confidence limit is required in the final results from that corresponding to the value of p used in (12.12).

For example, we can determine σ_{x_1} by obtaining Δx_1^S by (12.53) and putting

$$\sigma_{x_1} = \Delta x_1^S / p$$

(see (12.12) and (12.51)). And hence we can determine from normal distribution tables what multiple to take of σ_{x_1} for any required confidence limit, on the assumption that x_1^* is normally distributed (see (12.48)).

Further, a little consideration will show that the statistical approach is generally most effective for systems of high order, i.e., the ratios Δx_k^S to Δx_k^{II} are generally small when the number n of equations in the system is large.

12.9 Other Probability Distributions

The normal distribution is, of course, not the only one that can occur in practice, and we begin by considering probability distributions in general.

Given a distribution curve $y = \phi(x)$ the *mean* is defined by

$$\mu = \int_{-\infty}^{\infty} x\phi(x)\, dx \tag{12.59}$$

(compare with (12.2)), it being assumed that the area under the curve is unity, i.e.,

$$\int_{-\infty}^{\infty} \phi(x)\, dx = 1 \tag{12.60}$$

(see (12.5)). And the *variance* σ^2 is defined by

$$\sigma^2 = \int_{-\infty}^{\infty} (x - \mu)^2 \phi(x) \, dx \tag{12.61}$$

(compare with (12.4)).

Thus instead of defining the normal distribution as in (12.6) we might have defined it by

$$y = \frac{1}{\alpha(2\pi)^{\frac{1}{2}}} e^{-\frac{1}{2}(x-\beta)^2/\alpha^2}, \tag{12.62}$$

where α and β are parameters. Then, we could show by (12.59) and (12.61) that $\mu = \beta$ and $\sigma = \alpha$, so our definition would be equivalent to the one in (12.6).

Now, by way of an example of a distribution different from the normal distribution, let us suppose we are given a system of equations in which the coefficients and right-hand constants are known to a large number of decimal places. Then if such a system of equations is rounded to a small number d of decimal places the error introduced is not greater than $\varepsilon = 0 \cdot 5 \times 10^{-d}$. And the distribution curve for the uncertainty in each coefficient and constant may be taken to be of rectangular form as in Fig. 12.3.

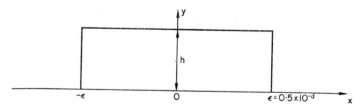

FIG. 12.3 Distribution curve for error when a random variable known to a large number of decimal places is rounded to d decimal places where d is small, the uncertainty ε being taken equal to the maximum possible error

Since the area under the curve in Fig. 12.3 is unity, by (12.60) we have that $2h\varepsilon = 1$ where h is the height of the rectangle, so that $h = 1/2\varepsilon$. Thus, the distribution curve for the error in each coefficient and constant is

$$y = \phi(x) = 1/2\varepsilon \qquad \text{for} \quad |x| \le \varepsilon$$

and
$$\tag{12.63}$$

$$y = \phi(x) = 0 \qquad \text{for} \quad |x| > \varepsilon.$$

The mean is clearly zero, i.e., $\mu = 0$. And the variance is given by

$$
\begin{aligned}
\sigma^2 &= \int_{-\infty}^{\infty} (x - \mu)^2 \phi(x) \, dx \\
&= \int_{-\varepsilon}^{\varepsilon} (x^2/2\varepsilon) \, dx = \frac{1}{2\varepsilon} \left[\frac{x^3}{3}\right]_{-\varepsilon}^{\varepsilon} = \varepsilon^2/3.
\end{aligned}
\tag{12.64}
$$

Hence, the standard deviation $\sigma = \varepsilon/\sqrt{3}$ so that

$$
\varepsilon = 3^{\frac{1}{2}}\sigma. \tag{12.65}
$$

Thus, taking the uncertainty ε for the distribution in Fig. 12.3 equal to the maximum possible error, corresponding to a confidence limit of 100 per cent, is equivalent to taking ε given by (12.65). Our choice of ε is thus equivalent to choosing $p = \sqrt{3}$ in (12.12).

Hence, with the Δx_k^{S} obtained by (12.53) or (12.56), it follows by (12.51) that

$$
\Delta x_k^{\mathrm{S}} = \sqrt{(3)}\, \sigma_{x_k}, \qquad k = 1, 2, \ldots, n. \tag{12.66}
$$

Suppose, now, that a confidence limit of 99·73 per cent is set for the errors in the solution not to exceed the uncertainties. For the normal distribution curve this requires taking each uncertainty equal to three times the corresponding standard deviation, it being noted that δx_k for each of $k = 1, 2, \ldots, n$ may be taken as normally distributed by the central limit theorem.

Thus, taking an interval with

$$
3\sigma_{x_k} = \sqrt{(3)}\, \Delta x_k^{\mathrm{S}}
$$

on either side of x_k, the statement

$$
x_k^* \in [x_k - \sqrt{(3)}\, \Delta x_k^{\mathrm{S}}, \, x_k + \sqrt{(3)}\, \Delta x_k^{\mathrm{S}}] \tag{12.67}
$$

holds with the required confidence limit of 99·73 per cent for each of $k = 1, 2, \ldots, n$.

Numerical Example

Let us now refer to the example in (4.36) and assume that the δx_k are each normally distributed, although each δx_k forms the sum of only a small number of independent random variables.

Then, since on rounding (4.36) to two decimal places we obtain (1.43), we now take the uncertainty ε in (1.44) to be 0·005 instead of 0·01.

Then, by analogy with (4.35) and (4.37) we have from (12.57) and (12.56) that

$$\Delta x_1^S = 0\cdot05379 \times (0\cdot005/0\cdot01) = 0\cdot02690$$
$$\Delta x_2^S = 0\cdot02720 \times (0\cdot005/0\cdot01) = 0\cdot01360 \qquad (12.68)$$
$$\Delta x_3^S = 0\cdot07070 \times (0\cdot005/0\cdot01) = 0\cdot03535.$$

And then, by (12.67), we have to multiply each of the statistical uncertainties in (12.68) by $\sqrt{3}$ to obtain uncertainties for which the confidence limit is 99·73 per cent:

$$\sqrt{(3)}\,\Delta x_1^S = 0\cdot04658, \qquad \sqrt{(3)}\,\Delta x_2^S = 0\cdot02356,$$
$$\sqrt{(3)}\,\Delta x_3^S = 0\cdot06123. \qquad (12.69)$$

Firstly, we note that these values are less than the uncertainties by Method II for $\varepsilon = 0\cdot005$ as given in (4.37):

$$\Delta x_1^{II} = 0\cdot0713, \qquad \Delta x_2^{II} = 0\cdot0348, \qquad \Delta x_3^{II} = 0\cdot1096, \qquad (12.70)$$

the ratios of the values in (12.69) to the corresponding values in (12.70) being

$$0\cdot6533, \qquad\qquad 0\cdot6770, \qquad\qquad 0\cdot5867. \qquad (12.71)$$

And, secondly, we note the actual errors given in (4.39)

$$|\delta x_1| = 0\cdot0270, \qquad |\delta x_2| = 0\cdot0038, \qquad |\delta x_3| = 0\cdot0438 \qquad (12.72)$$

are well within the estimates obtained by statistical considerations for a confidence limit of 99·73 per cent (see (12.69)).

Clearly, the value of the statistical approach is illustrated in this example, too.

12.10 Another Practical Situation

In Sections 12.4 to 12.8, we considered the case which required a knowledge of the distribution curve for the true values (see (12.8) and (12.9)).

In that situation, we may have had to approximate the population mean and population standard deviation by the sample mean and sample standard deviation. But, of course, the value of a sample measure depends on the sample. And we note (see reference 1, pp. 50 and 76) that:

> The distribution curve for $\bar{x}$, the mean of a random sample of n measurements, has mean μ and standard deviation $\sigma/\sqrt{n}$ where μ and σ refer to the corresponding population measures. $\qquad (12.73)$

and that:

> The distribution curve of the standard deviation s of a
> random sample of n measurements can be obtained
> from the fact that $(n - 1)s^2/\sigma^2$ is a random variable   (12.74)
> having a chi-square distribution with $n - 1$ degrees of
> freedom.

Now the first of the two results follows easily from (12.24) and
(12.25). For $\bar{x}$, the mean of n measurements $m_1, m_2, \ldots, m_n$, is a random
variable, being a linear combination of the n independent random
variables $m_1, m_2, \ldots, m_n$:

$$\bar{x} = \frac{1}{n}m_1 + \frac{1}{n}m_2 + \ldots + \frac{1}{n}m_n. \qquad (12.75)$$

And it follows by the central limit theorem stated in Section 12.4 that:

> The sample mean of a random variable is normally
> distributed if the random variable is itself normally
> distributed but otherwise nevertheless approaches the   (12.76)
> normal distribution, the sample mean in our treatment
> being taken to be normally distributed.

Now, the theory in Sections 12.4 to 12.8 could have been developed
using the results in (12.73), (12.74), and (12.76), instead of merely
approximating μ and σ by $\bar{x}$ and s. But the approximations may be good
enough in practice, especially where the number of observations n is
not small. It will be seen, however, that in the present section we have
to use the results in (12.73) and (12.76) although we still conveniently
approximate σ by s without using (12.74).

Now, the practical situation dealt with in this section is where a
number of measurements or observations are made for each coefficient
and right-hand constant. In this context, the term 'true value', then,
refers to the population mean of the measurements.

Our procedure is to take the approximate values of the coefficients
and right-hand constants to be given by the means of the measurements
for them, this approach now not being a matter of approximation. And
the intervals of uncertainty must now be constructed in such a way as to
include μ for the confidence limit in question, because μ is our true
value now.

Now, there is a 99·73 per cent confidence limit that $\bar{x}$ lies in the
interval in

$$[\mu - 3\sigma/\sqrt{n}, \qquad \mu + 3\sigma/\sqrt{n}], \qquad (12.77)$$

$\bar{x}$ being normally distributed by (12.76) (see Fig. 12.4). But when $\bar{x}$ lies in the interval in (12.77) it is clear that the interval of minimum width that certainly contains μ is

$$[\bar{x} - 3\sigma/\sqrt{n}, \quad \bar{x} + 3\sigma/\sqrt{n}], \tag{12.78}$$

i.e.,

$$\mu \in [\bar{x} - 3\sigma/\sqrt{n}, \quad \bar{x} + 3\sigma/\sqrt{n}]. \tag{12.79}$$

Hence, the interval of uncertainty in (12.78) can be given with a confidence limit of 99·73 per cent for it to contain the true value.

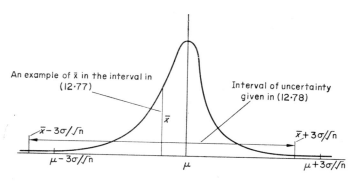

FIG. 12.4   Distribution curve of the mean $\bar{x}$ of a random sample of n observations, the mean $\bar{x}$ being a random variable with mean μ and standard deviation $\sigma/\sqrt{n}$ where μ and σ refer to the original population

In general, an interval of uncertainty given by

$$[\bar{x} - p\sigma/\sqrt{n}, \bar{x} + p\sigma/\sqrt{n}] \tag{12.80}$$

in place of the one in (12.78) will ensure that the interval will contain μ with a confidence limit specified by p for a normal distribution curve (see (12.10) and (12.76)).

By way of notation, let us now suppose that n_{ij} measurements are made for a_{ij}^*, $i, j = 1, 2, \ldots, n$, and n_i measurements for c_i^*, $i = 1, 2, \ldots, n$. Then the standard deviation for the means are

$$\bar{\sigma}_{ij} = \sigma_{ij}/\sqrt{n_{ij}}, \quad \bar{\sigma}_i = \sigma_i/\sqrt{n_i}, \quad i, j = 1, 2, \ldots, n, \tag{12.81}$$

where σ_{ij} and σ_i are the standard deviations of the original populations from which the random samples are chosen. And corresponding to (12.12) we now have

$$\varepsilon_{ij} = p\bar{\sigma}_{ij} = p\sigma_{ij}/\sqrt{n_{ij}}, \quad \varepsilon_i = p\bar{\sigma}_i = p\sigma_i/\sqrt{n_i},$$
$$i, j = 1, 2, \ldots, n. \tag{12.82}$$

Now, as indicated earlier, we take the sample means to represent the approximate values a_{ij} and c_i in this section. Hence, the confidence limit for the following statements to be true

$$a_{ij}^* \in [a_{ij} - \varepsilon_{ij}, a_{ij} + \varepsilon_{ij}], \qquad i, j = 1, 2, \ldots, n,$$
$$c_i^* \in [c_i - \varepsilon_i, c_i + \varepsilon_i], \qquad i = 1, 2, \ldots, n, \tag{12.83}$$

clearly depends on our choice of p in (12.80) or (12.82), the true values a_{ij}^* and c_i^* now being the population means of the observations (see (12.77) to (12.80) and also (12.13)).

And a little consideration will show that the quantities

$$\delta a_{ij} = a_{ij} - a_{ij}^* \qquad \text{and} \qquad \delta c_i = c_i - c_i^* \tag{12.84}$$

are random variables with zero means and standard deviations $\bar{\sigma}_{ij}$ and $\bar{\sigma}_i$. For, in the present context, the a_{ij} and c_i can be regarded as random variables with population means a_{ij}^* and c_i^* and standard deviations $\bar{\sigma}_{ij}$ and $\bar{\sigma}_i$ (see (12.18) to (12.21) and (12.73)).

Hence, the development of the theory now proceeds as before with (12.82) to (12.84) in place of (12.12), (12.13), (12.18), and (12.20). And the results have the same form except that whenever σ_{ij} or σ_i appeared previously we now have $\bar{\sigma}_{ij} = \sigma_{ij}/\sqrt{n_{ij}}$ or $\bar{\sigma}_i = \sigma_i/\sqrt{n_i}$, the σ_{ii} and σ_i referring to the original populations. In particular, in place of (12.32), (12.41), and (12.49), we now have

$$\sigma_a^2 = a^2 \sum_{i=1}^{n} \sum_{j=1}^{n} b_{ji}^2 \bar{\sigma}_{ij}^2 = a^2 \sum_{i=1}^{n} \sum_{j=1}^{n} b_{ji}^2 \sigma_{ij}^2 / n_{ij} \tag{12.85}$$

$$\phi_s = p \left(\sum_{i=1}^{n} \sum_{j=1}^{n} b_{ji}^2 \bar{\sigma}_{ij}^2 \right)^{\frac{1}{2}} = p \left(\sum_{i=1}^{n} \sum_{j=1}^{n} b_{ji}^2 \sigma_{ij}^2 / n_{ij} \right)^{\frac{1}{2}} \tag{12.86}$$

$$\sigma_{x_k} = \left(\sum_{i=1}^{n} \sum_{j=1}^{n} (x_j b_{ki})^2 \bar{\sigma}_{ij}^2 + \sum_{i=1}^{n} b_{ki}^2 \bar{\sigma}_i^2 \right)^{\frac{1}{2}}$$

$$= \left(\sum_{i=1}^{n} \sum_{j=1}^{n} (x_j b_{ki})^2 \sigma_{ij}^2 / n_{ij} + \sum_{i=1}^{n} b_{ki}^2 \sigma_i^2 / n_i \right)^{\frac{1}{2}}, \quad k = 1, 2, \ldots, n. \tag{12.87}$$

In practice, it may still be convenient to approximate σ_{ij} and σ_i by their sample values, although the distribution curves for the σ_{ij} and σ_i could be taken into account (see (12.74)).

We have thus dealt in Sections 12.4 to 12.8, 12.9, and in the present section with the statistical approach to the approximate equation problem in different situations. In all cases, we have to decide what values to assign to the approximate coefficients a_{ij} and right-hand constants c_i

and how to construct the intervals of uncertainty for given confidence limits.

In concluding this chapter, we may say that by and large the statistical approach is very useful and leads to smaller and more meaningful values for the uncertainties. And the statistical approach can be expected to be particularly effective when n, the number of equations in the system, is large.

As regards which is the best, most practical, or recommended method for the solution of the approximate equation problem we refer the reader to the remarks in the preface to this book.

Reference

1. BOWKER, A. H. and LIEBERMAN, G. J. *Engineering Statistics*, Prentice-Hall, Englewood Cliffs, N.J. (1959).

Index

221